PENGUIN BOOKS

Kings of the Yukon

'Weymouth combines acute political, personal and ecological understanding, with the most beautiful writing reminiscent of a young Robert Macfarlane . . . He is, I have no doubt, a significant voice for the future . . . a really outstanding new contemporary British voice . . . I've never seen such a strong and excited consensus among the judges for a winner' Andrew Holgate, *Sunday Times* literary editor and judge of the *Sunday Times* Young Writer of the Year Award 2018

'This is the best kind of travel writing. Weymouth embarks on an ambitious journey – 2,000 miles down the Yukon in a canoe – voyaging, listening and learning. An outstanding book'
Rob Penn, author of *The Man Who Made Things Out of Trees*

'Lyrical . . . The elegiac tone that fills *Kings of the Yukon*, the sorrow at the loss of culture and nature in the wilderness, is an unavoidable reflection of life in the 21st century' Richard Lea, *Guardian*

'An infatuated love letter to the river' Chris Fitch, *Geographical*

'Enthralling . . . vividly conveys the raw grandeur and deep silences of the Yukon landscape' Luke Jennings, author of *Blood Knots*

'Dazzling, often in unexpected ways, Adam Weymouth is a wonderful travel writer, nature writer, adventure writer – along the way, he is also a nuanced examiner of some of the world's most fraught and urgent questions about the interconnectedness of people and the natural world'
Kamila Shamsie, author of *Home Fire*

'Adam Weymouth paddled an 18ft glass-fibre canoe down the Yukon, almost 2,000 miles through Canada and Alaska, to the Bering Sea. His account of that journey is so assured, so accomplished, that I found it hard to believe it was his first book. He hoped he could explain the decline in numbers of king salmon and show how the lives of those who depend on the fish are changing. If he is frustrated in his first objective, he succeeds fully in his second, in his tracing of the relationship between fish and people, and "of the imprint that one leaves on the other". It's a story about a sparsely populated place but one that's rich in characters, and it's beautifully written'
Michael Kerr, *Daily Telegraph*, Books of the Year

'A brilliant account of a summer spent paddling the 2,000-mile length of the Yukon River . . . Kings of the Yukon succeeds as an adventure tale, a natural history and a work of art. Its various threads of context and back story are woven seamlessly into the daily panorama of the river journey'
Richard Adams Carey, *Wall Street Journal*

'Adam Weymouth's account of his canoe trip down the Yukon River is both stirring and heartbreaking. He ably describes a world that seems alternately untouched by human beings and teetering at the brink of ruin'
David Owen, author of *Where the Water Goes*

'A moving, masterful portrait of a river, the people who live on its banks, and the salmon that connect their lives to the land. It is at once travelogue, natural history, and a meditation on the sort of wildness of which we are intrinsically a part. Adam Weymouth deftly illuminates the symbiosis between humans and the natural world – a relationship so ancient, complex, and mysterious that it just might save us' Kate Harris, author of *Lands of Lost Borders*

'Shift over Pierre Berton and Farley Mowat. You, too, Robert Service. Set another place at the table for Adam Weymouth, who writes as powerfully and poetically about the Far North as any of the greats who went before him'
Roy MacGregor, author of *Original Highways: Travelling the Great Rivers of Canada*

'Adam Weymouth writes of the Yukon River, the salmon and the people, with language that flows and ripples like the water he describes. There may be a smoothness to the words, but pay attention, there are deep undercurrents here. You can hear the water dripping from his paddle between each stroke as he travels that river. It mingles with the voices of the many people he visits along its shores' Harold Johnson, author of *Firewater: How Alcohol Is Killing My People (and Yours)*

ABOUT THE AUTHOR

Adam Weymouth's work has been published by a wide variety of outlets including the *Guardian*, the *Atlantic* and the *New Internationalist*. His interest in the relationship between humans and the world around them has led him to write on issues of climate change and environmentalism, and most recently, to the Yukon river and the stories of the communities living on its banks. He lives on a 100-year-old Dutch barge on the River Lea in London. This is his first book.

Kings of the Yukon

An Alaskan River Journey

ADAM WEYMOUTH

PENGUIN BOOKS

PENGUIN BOOKS

UK | USA | Canada | Ireland | Australia
India | New Zealand | South Africa

Penguin Books is part of the Penguin Random House group of companies whose addresses can be found at global.penguinrandomhouse.com.

First published by Particular Books 2018
Published in Penguin Books 2019
001

Illustration and map by Ulli Mattsson

Set in 11.4/14 pt BemboBookMTStd
Typeset by Jouve (UK), Milton Keynes
Printed and bound in Great Britain by Clays Ltd, Elcograf S.p.A.

A CIP catalogue record for this book is available from the British Library

ISBN: 978–0–141–98379–0

www.greenpenguin.co.uk

To you, as yet unnamed, who came along quicker than this book

And to Mum and Dad, for always trusting that I knew what I was doing

– the generations that came before, the generations coming after

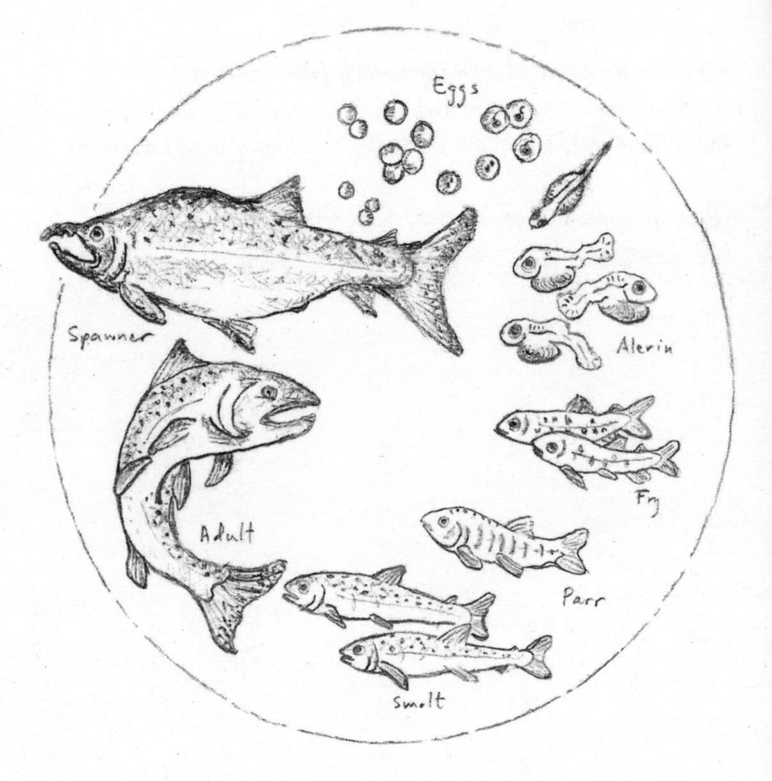
Eggs
Alevin
Fry
Parr
Smolt
Adult
Spawner

Wait, I see something: We come upstream in red canoes.
Riddle of the Alaskan Athabascans, documented by the missionary Julius Jette (1864–1927)

rival (n.)
from the Latin *rivalis*, 'one person using the same stream as another'

Contents

Map x
Author's Note xii

Kings of the Yukon 1

Acknowledgements 265

RUSSIA
Emmonak
YUKON
Holy Cross
Pilot Station
Bethel
N
BERING SEA

BOUNDARY
LINE
CANADA
Porcupine R.
ARCTIC CIRCLE
Fort Yukon
RIVER
Jo's Fishcamp
Circle
Eagle
Rock Island
Fairbanks
Dawson City
McNeil R.
Tatchun Creek
Carmacks
White R.
Whitehorse
Teslin R.
Nisutlin R.
Marsh Lake
Teslin Lake
Teslin
Atlin
Mt. St. Elias
NORTH PACIFIC OCEAN

Author's Note

There are five species of salmon found in Alaska and Western Canada. This book is concerned predominantly with one of them, *Oncorhynchus tshawytscha*, generally referred to as the king in Alaska and the Chinook in Canada. I have used these names interchangeably throughout.

Whilst the majority of the research for *Kings of the Yukon* was carried out during the summer of 2016, I returned to the Canadian side of the border for a shorter trip in 2017. One summer in the North is simply too short a season to cover the two thousand miles of river that the salmon span between break-up and freeze-up. I have, however, written about the trip as one continuous journey, in order to preserve the flow of the story. All data on escapement numbers, sex ratios, etc., relates to the 2016 season. Most interviews were recorded by hand, either at the time or afterwards; several were recorded on audio. Some characters' names have been changed to protect their privacy. Interviews in Canada were carried out according to the Traditional Knowledge policies of the First Nations involved, policies which strive to protect their cultural heritage from exploitation.

The terms 'Eskimo' and 'Indian' are often considered pejorative, yet are commonly in use in Alaska and Canada, amongst both Indigenous and white people. 'Alaskan Native' and 'First Nation' are not able to differentiate between these two very separate groups, and they do not encompass the connection, on the Eskimo side, between other groups that inhabit the circumpolar region, and on the Indian side, with other native peoples of Canada and the Lower 48. As such I have occasionally used them in the book, alongside the proper names of specific tribes and clans. Whilst the word 'Indian' is commonly assumed to derive from Columbus believing that he had reached India, figures such as the activist Russell Means and the American Indian Movement offer an alternative etymology, described here by Cree lawyer Harold R. Johnson in his book *Firewater*: 'Columbus was not lost, he knew where he was, and he called us *In Dios*, meaning "with God". The word is not as important as the story we tell about it. Indian is also a precise legal term found in our Treaties and the Canadian constitution.'

Finally, for reasons best known to themselves, Alaskans call a snowmobile a snowmachine. A snowmachine is not for making snow, as it is everywhere else in the world, but for driving on it. It is the word I have used here.

Kings of the Yukon

The river is flowing backwards, back up from the sea.

They swim through silt, eyes wide, unblinking. Thirty, forty, fifty pounds of flesh, many thousands of them. Their backs speckled like frogspawn, the blush of their bellies, where the silver of their flanks fades into a deep and meaty rose. Jaws gawping, lips beginning to curve in upon themselves like pliers, propping their mouths ajar so that the river flows right through them, and yet for the rest of their lives these salmon will not eat, they will not drink. The organs that sustain them, the kidneys and the stomach, are shrinking as they sense the sweet water they have not known since they were smolts, finger big, years back. Familiar scents long forgotten are triggering changes in their brains and in their bodies: their chemistry has new priorities now. The ovaries of the female hen will swell to a sixth of her bodyweight during her swim upriver, the cock's testes will increase fivefold.

For several years the salmon have roamed in schools throughout the Bering Sea, the chain of the Aleutians, the Sea of Okhotsk, ranging as far as the coasts of Japan. Their many species mingle. They travel further than science can reach, and much of what they eat and where they winter can only be surmised. The Yup'ik say that they live in human form beneath the waves, five houses, one for each tribe, and at the salmon king's behest, each spring, they pull on their fins and silver skins and make for the human world. In the late Arctic spring, impelled before the others, the kings turn for North America's west coast. California, Oregon, Washington, British Columbia, Alaska. They have iron in their brains, their compasses have led them here; now they scent their birthing pools. A chemical mix of vegetable and mineral unique to the waters of their birth that draws them on a thread up several thousand miles of river. They can distinguish a single drop from their home river amongst two million gallons of seawater.

The movement of one, changing direction, pulses through the rest, electric. At times they crest the surface, a dark sleek back, a dorsal fin, undulating like dolphins. These fish are many pounds of muscle, toned through years of swimming headlong into Pacific storms, and their flesh is red as blood. They force against the Yukon's current, shouldering their way upriver, tacking crosswise through the flow, setting their fins like sails. Their shadows pass like clouds across the bottom. They rest in the eddies of boulders on the river's bed, erratics left behind by glaciers. At the river's mouth the water is still brackish, the current compounded with the flow of the tide. But it is diluting, and as they move further up the delta the sea slackens off its hold, resigning, letting them go. On this great in-breath of the land the kings spread up through the waters and their tributaries, permeating the watershed. Eventually, they will push thousands of miles into the continent's interior. They will reach mountain lakes, they will reach the clouds.

It is the end of May, and spring is late. Which is not to say that it is a late spring, because this far north there is no such thing as a yardstick to measure the seasons up against ('I've been here fifty years,' in the words of one old-timer, 'and the only normal year we had was two years ago'), but people say it is not where it ought to be. Mountains ring the town, and the snow still comes far down their peaks. The dandelions only opened last week. Bears have awoken from hibernation and, finding nothing to eat, have roamed within the Whitehorse city limits, sniffing out trashcans and chicken coops. The government has shot six in the past month.

Whitehorse, in the Yukon Territory, northwest Canada, is so named for the white manes that once tossed in the rapids of Miles Canyon, but that was more than a hundred years ago, back in Gold Rush times, almost prehistory; the dam went up in '58 and the horses are all since drowned. Above the dam is a lake called Schwatka, after the man who mapped and did his best to tame the Yukon River on behalf of the US Army, and where Yukoners go to test drive their jet skis come the summertime. The floatplane base is little more than a jetty of planks buoyed up by oil drums, an old-fashioned balance to weigh in the bags, and a Cessna plane moored up on a thin rope, bright red against the blue. The pilot has only recently switched his skis over to floats. He cannot tell us whether McNeil Lake, where we are headed, will still be frozen over. I was in London three days ago and I am not used to uncertainty. But up here, out there, if someone hasn't seen it then the information doesn't exist.

Our bags come in at just below the permitted eight hundred pounds. The pilot asks our weights, so as best to arrange us in the plane. It is a perfect day, not a breath, and the mirrored clouds sit, precisely, upon the surface of the lake. Inside the cockpit it is sweltering. Take-off can be complicated when the surface is so glassy, no friction to catch against.

'Going to be a fun one,' grins the pilot, in cap and Ray Bans and check shirt, tucked-in, hauling up the bags, but before this he was out in Afghanistan and I assume that he lives for the fun ones. We strap ourselves in and put on the headsets and chug up to the inlet of the lake to maximize the length for take-off. He cuts the engine. We bob around. The lap of water, lolling against the floats.

I am travelling with Hector MacKenzie, a Scot who moved to Canada back when it was easy to move to Canada, as a young man in the late '60s. He came and liked what he saw and stayed.

He spent many years out in the bush, before moving into Whitehorse when his kids hit an age where they required more by way of schooling than a homestead could provide. Hector has climbed and guided and paddled all over the world. Now retired and in his seventies, his life is little different, except that he stays closer to home and canoes and skis only when the fancy takes him, which is often. A trim beard, white hair, a face that has stared down much weather. He exudes calm, which I welcome: up until now, I have spent perhaps a week in a canoe, on British rivers that look, to Canadian eyes, like trickles – the Medway, the Dart, the Wye. I have learnt my paddle strokes from books.

'We're just getting a bit of a northwesterly now, giving you a south-north departure,' says the radio.

'That's us,' the pilot says.

He starts the motor, and eases up the throttle. We begin to plane across the lake's flat surface, spray flying from the floats. He fiddles dials and plugs something into the navigational display. We race past a couple of fishermen, out in a boat with rod and reel, their hands raised. The end of the lake, where the dam begins, is rapidly approaching.

'Some pilots try and take off right away,' he says over the intercom. 'You can't pretend you're flying. You just treat her like a speedboat and she'll come up.' He nods. 'You feel the heels come up there?'

I look back through the windows and there is no longer spray. The pilot pulls back hard, and we rise up above the dam and up into the sky.

We climb. I look north, and already I cannot make out Hector's house on the outskirts of the city. Whitehorse is a one-horse town, capital of the Yukon Territory, population twenty-five thousand. There are only thirty-five thousand people in the territory. People are fond of saying that there are more moose than people in the Yukon Territory, but in reality there are more of most things: more beaver, more salmon, more square miles. The red rust stain of a molybdenum mine grazes one slope, but for the rest it is valleys and forests of spruce, and the occasional trail cut through them.

Whitehorse's first boom came in the last decade of the nineteenth century with the construction of the railroad, a link from coast to city, from where a sternwheeler to Dawson City made for a safe, if expensive, route to the Klondike goldfields. The second boom came with the construction of the Alaska Highway in 1942, two lanes that connected Alaska to Whitehorse, at the Canadian border, and from there to Vancouver and to the rest of the world. The population doubled, a third of the residents were squatters, the average age was about right to buy a beer. An old lady in town told me that when she first took her young kids south to Vancouver they cried because they had never seen anybody with wrinkles. There are many who came that never left.

We fly for an hour, heading east. Below, our shadow trails us. The lakes are as big as inland seas. Those in shade at the basins of summits are still frozen or still thawing, blue and white, like marbled paper. But up ahead Moss Lake is open, and that is a good sign, being as it is only a hundred metres lower than McNeil Lake, where we are headed. We cross rivers that run thin and white in sunlight. We pass Mount Hogg, Mount Placid, the mountains of the Big Salmon Range, and all the peaks that have never been named. We see no animals.

'Do you come up this way much?' I ask the pilot through the headset.

'Summertime I bring up hunters,' he replies.

'Moose?'

He nods. 'Moose, sheep, goat, bear. Anything that walks people will shoot it.' He asks me what I am doing up here. I tell him I am looking for king salmon.

*

There are five species of Pacific salmon in North America: the chum, the coho, the sockeye, the pink, and the Chinook. Each has its own diminutive: the chum is the dog, or the keta, the coho the silver, the sockeye the red, the pink the humpy, and the Chinook is the king. The original Chinook are people of the Pacific Northwest, and their language formed the core of Chinook Jargon, a pidgin trading language that stretched from Alaska to the Columbia River, along what now forms the border of Washington and Oregon, and incorporated the words of many tribes, as well as French and English. Any Canadian will still say Chinook for king, the best and biggest of the fish that the Chinook people traded.

The Pacific salmon (*Oncorhynchus*, Greek for hook nose) and the Atlantic salmon (*Salmo salar*) share a common ancestor. They diverged between twenty and fifteen million years ago, during the Miocene cooling of the Arctic Ocean, which put between them a barrier of water too cold to cross. Separated, they became two distinct genera, but unlike the Atlantic salmon, which did not split further, the Pacific salmon went on diverging, resulting in the several distinct species of today (via some other remarkable examples: *Smilodonichthys rastrosus*, the sabre-toothed salmon, 10 feet long and 350 pounds). Each of these species has found its niche in the lakes, estuaries and mountain creeks of North America's west coast. It is widely believed that *Oncorhynchus* was shaped by the dramatic geological changes along this coast, compared to the more stable east: huge uplifts along the Pacific Rim that produced the Sierras, the Rockies, the Alaska Range, all crushed into the sky, creating the varied habitats that allowed for the evolution of five species. And these species mutated further, each individual population adapting to the spawning ground it occupied, resulting in populations as genetically diverse as the number of spawning streams. Restless, migratory, wide-ranging across the oceans, they have returned in successive generations to the exact same streams where they were birthed, mother followed by daughter, father followed by son, for many tens of thousands of years. Pilgrimages are thought to have begun with nomads going back to the graves of their ancestors. Such is the salmon's return.

The history of the salmon is the history of this land. Rock carvings in Alaska at the mouths of salmon streams are ten thousand years old. Big salmon etched into sandstone. Small salmon swimming upstream. A two-headed salmon, which suggests an awareness of their life cycle, of their departure and return. Carved by shamans, perhaps, to invoke the salmon's annual pilgrimage; or insignia to mark the fishing rights to different salmon streams; or doodles, drawn on the slow days when the runs didn't show or the river was in spate. Once people here would have had nets made from the sinews of rabbits and babiche, fish traps as funnels of woven willow. Dip nets made from willow bark, and spears tipped with bone. Chum salmon bones found in middens in the Tanana River Valley have been dated back eleven and a half thousand years.

The Yukon River is the longest salmon run in the world. Where exactly the Yukon has its source will never be resolved because there is no single answer, with countless tributaries rising across western Canada. There are 110 known rivers where the king spawns in the Yukon Territory alone, but the McNeil, where we are headed, is the furthest of those that the kings are known to reach. No species goes further, which is to say, the few kings that make it back to McNeil Lake have travelled further up a river than any other salmon on the planet.

These kings arrive at the Yukon's mouth from somewhere deep in the Pacific, and they swim upriver, against the current, on a path that bisects Alaska, crossing the border into Canada's Yukon Territory, taking a left up the Teslin River, crossing Teslin Lake, entering the Nisutlin River at the inlet to Nisutlin

Bay, swimming through Moss Lake, and then up the McNeil River to its source. From the plane we can see some of that map, as far off as the Teslin, eighty miles away. The McNeil River is beneath us now, meandering across its floodplain, and as we crest the far end of its valley we see the lake and it is turquoise, sparkling, as though the several feet of ice that cover it for half the year were no more than a dream.

'Nice spot,' the pilot says. We fly up the lake's eastern shore and bank and touch down on the water.

The ridge at the northern end of McNeil Lake, the St Cyr Range, its broad summits piebald with snow, is the traditional dividing line between the Tlingit and the Dene people, between the coastal and the inland, between the watersheds of the Yukon and the MacKenzie, and it is where the kings come to a stop. They are capable of herculean feats, but are unable to haul themselves on their fins over the crest of the ridge and gush down the other side. The ones that make it to the lake have climbed 1,054 metres into the mountains, and have swum against the river's current for 1,990 miles. Just this week, on the daily email briefing from Alaska's Department of Fish and Game (ADF&G), I read that the first of the year's kings have been seen at the Yukon's mouth, coming up through the tides. One or two of them will be bound for here, by forces as undeniable as gravity. By the time they arrive, at the beginning of September, I am planning to be down at the mouth. Somewhere along the way, our paths will surely cross.

The pilot leaves Hector and me on a gravel shore. It is cold, and the clouds are rolling in. We watch as the plane rises back into the sky, our hands raised high, and turns back towards Whitehorse. Then there is silence. We have a mound of stuff with us. Fishing rods and pots and pans and chairs and dry suits and clothes and medicines and ropes and a deflated raft and ratchets and paddles and spare paddles and four weeks' worth of food, or thereabouts, and an enormous canvas tent. We are beside an empty trapper's cabin, ply-nailed across the windows to keep away the bears. Rusted traps hang on pegs outside. It is the only sign of human life for perhaps a hundred miles. A shallow creek, with a stony bed, meets the lake here. A couple of sandpipers on stubby wings skim the surface of the stream. Their trilling whistle, which I hear for the first time, will become the melody of the river.

We camp a few hundred metres down the shore, away from the cabin, beside the outlet of the lake where the current starts to gather. It is getting late, and colder, but it will not be dark until, well, August, and we are not in any rush. Hector sets the tent and I gather wood, a mix of drift from along the shore and resinous dead branches pulled from the bases of the spruce trees. We cook a supper of rice and vegetables, and make a start on the meat that will not keep long. We have some whiskey that we can eke out for a few weeks if we are careful. After we eat, I sit outside the tent beside the fire. The channel is narrow enough that you could throw a rock to the far bank. The water draws your eye, far more than any fire does; it would feel odd to have one's back to it. Passing, passing, already urgent for the sea. It will be there long before I am. You stare for long enough, and when you raise your eyes it is the land that seems to move.

It is as though we have gone back in time, back in season, to just after winter broke. There is scarcely any new growth here. The catkins of the willows are tightly furled, and besides the willow there are no trees but spruce this close to the tree line. Sheets of ice bob in the sloughs, the side channels that branch out from the river. Only one species of flower, speckling the beaches, a sort of anemone, five white petals and a blue tinge underneath, like fine weather seen through clouds.

Three days, I think, from the centre of London. That's all it takes.

*

I wake with a start; I must have overslept. Light streams in through the tent. I grab my watch. One a.m. I stare at it, uncomprehending. I fall back to sleep again.

I first came to the North in 2013. I came, ostensibly, to follow stories of climate change and oil, but really I came because I was lured in by the myth. I had grown up on Jack London and Farley Mowat; I had fleshed it out with Jon Krakauer's *Into the Wild* and bluegrass and wildlife documentaries. I spent three months visiting every corner that I could: the site of the Exxon Valdez oil spill, twenty-five years on; a whaling festival in the Inuit village of Point Hope, the oldest settled site in North America; Newtok, slipping away into the sea as the permafrost beneath it thawed; Chicken, where gold miners from across the state came to celebrate the Fourth of July. I had read about a man called Mike Williams, chief of the Yupiit nation, who had spoken widely about the climatic changes that his people were experiencing. He invited me to his village, way out on the Yukon-Kuskokwim Delta. All flights to the delta pass through Bethel, and it so happened that whilst I was in town Mike was coordinating the trial of twenty-three Yup'ik fishermen who were in court for catching king salmon during a fishing closure in the summer of 2012.

'Gandhi had his salt, we have our salmon,' Mike said.

The closure had been implemented by ADF&G as king salmon numbers plummeted, unexpectedly and inexplicably. The fishermen pleaded not guilty. They were justified in fishing, so they said, because the taking of king salmon was part of their spiritual practice, their cultural heritage.

It was not big news; it made the local radio. But on the delta it was significant. There were tears in the courtroom. I wrote an article, which ran in *The Atlantic*. Back at home, I kept an eye on events. In 2014 a ban on all fishing for king salmon, commercial and subsistence, was enacted along the entire Yukon River, Alaskan and Canadian, an unprecedented move. In 2015 it was kept in place, albeit with some minimal harvest on the Canadian side of the border. In Canada the self-governing First Nations determine their own fishing quotas, with the national government only permitted to step in if a situation becomes critical; in the United States regulations are determined by the state departments of Fish and Game. Since the blanket bans of 2014 and 2015, ADF&G have been attempting to let people get some fish for themselves, whilst remaining conservation minded. That approach is in marked contrast to the years before the crash in salmon numbers, when the department was accused, from many sides, of driving the run into the ground through a lack of regulation.

No one is pretending that managing salmon is simple. Fred Andersen, a former fisheries biologist, has called the Yukon probably the most complex salmon fishery in the world. There are three thousand acronyms in play. Fisheries managers are attempting to regulate for a subsistence fishery, a commercial fishery, and a healthy ecosystem. During the fishing season, ADF&G will speak every single morning to discuss the strategy for the day. Stephanie Quinn-Davidson, who managed the Chinook run until 2015, told me that 85 per cent of her job was communication, with other managers, with media and, most crucially, with fishermen. The kings pass through a mosaic of state and federal land, and through the United States and Canada, each country with different divisions to its fisheries. Subsistence fishing is treated differently under state and federal law. Diplomacy between the United States and Canada is a further complication: the Yukon River Salmon Agreement, hammered out over thirty years, requires that fishing in Alaska be restricted to a point whereby between 42,500 and 55,000 kings be let across the border into Canada each year. This is a target that the US has failed to deliver five times in the past ten years, most recently in 2013.

In Alaska the runs of certain salmon species – the pink, the sockeye, the chum - still hit the millions on some rivers. Those on the outside forced up onto the banks. The backs of the ones above burnt by the sun, so they say, the bellies of the ones beneath scoured by the gravel, so thickly are they crowded. It is iconic, the ultimate symbol of the wild, the fishing trip to die for. Across the world, from the Far East to Europe, from North America's east coast to its west, rivers that once knew a similar abundance see a fraction of their historic numbers; many have lost their salmon altogether. The River Salm, in Germany, after which the salmon is thought to be named, has no salmon any more. At one time Alaska was nothing exceptional. Now it's simply all that's left.

Yet in everything I read no one seemed able to agree whether the declines had come from poor management, long-term climatic changes, bycatch in the oceans, disease, or natural fluctuations. What seemed certain was that not only did the future of the king hang in the balance on the Yukon, this was the last chance on earth to get it right. The king is Alaska's state fish, and across a border that the salmon do not recognize, in Western Canada, the bond to the Chinook is the same. If I was to try and better understand the North, I thought, then perhaps I should go looking for one of its most iconic species, the royalty of the river, before it was gone for good.

The Chinook has threaded together the communities that live along the Yukon for millennia. It intimately connected the lives of a Tlingit Indian at the river's source and a Yupik Eskimo on Alaska's coast, two thousand miles away, long before these people were aware of each other's existence. It is a link to peoples' ancestors and their hope for their children's children. Many of the Yukon River's villages, once the sites of seasonal fish camps, remain hundreds of miles from the closest road, and unless going by plane there is no way in but for the river.

Travelling by boat at the same time as the king run and covering the same path as the fish, albeit paddling in the opposite direction, I hoped to better understand what is changing, not just in the life of the river, but in the lives of the people that depend on it. And I wanted to see how one of the most remote regions on the planet is experiencing the climatic and economic forces that are shaping the rest of our world. I had made long, slow journeys across countries before, and had found this way of travelling revealing in the interactions that it opened up and the stories that people shared, the trust that people placed in me because I had come from a village that they knew, was heading somewhere they had been. Now I wondered whether such a journey could also shed light on what was happening to the Yukon's kings.

At half past six I wake once more. I am still jet lagged, yet to catch up with myself, and cannot fathom all this *light*. The river has been muttering all night. I step from the tent, and a pair of harlequin ducks, their faces painted, burst from the shore and fly, complaining, off down the meanders of the river. There is a skin of frost on all the gear. I make a small fire and brew some coffee in the little blackened kettle that has been everywhere with me. Mist hangs in the branches of the spruce on the far bank, and the air is chill and damp. I wear a coat and hat and gloves, but it is a chill that goes right through them. I put a pot of porridge on the embers, and watch as it bubbles.

The sand holds the stories of the night. The hoof prints of a moose and calf, like two series of quotation marks, emerging from the water and making off into the willows. The little nervous arrows of the sandpiper, pointing every which way, all along the shore. These female sandpipers are polyandrous: each male that she mates with she will leave to incubate her eggs whilst she goes off searching for another. Hector tells me this as we eat.

Hector loves birds; he speaks of them as though they were his own prodigious children. Later, as we lie in the tent one night, listening to the whistle of some bird, he will suddenly, urgently, whisper:

'Hear that? The semi-palmated plover. It is a sound that thrills me to my very marrow.'

I stand to do the dishes. There is an eddy here, on the outside bend, where the river begins in earnest. The water is slack and slowly turning, contrary to the flow, a deep pool slowly carved. I am scrubbing porridge from out the pot when I see the dart of little fish emerging from the shadows, pecking at the oats. And I gasp because it was as easy as that to find them. They are exactly where they were supposed to be. These are king salmon, a few months old, the first I have ever seen. I watch them as they feed on our breakfast.

At this stage, in the complicated lexicon of a fish that has spawned its own glossary, they are called parr. It is the point where the parr marks appear, the oblong splotches that run the length of them like inky fingerprints and, like fingerprints, distinguish one species from another. The marks are obscured as they age, but for now they are camouflage against the gulls and terns and pike and mergansers and otters and sheefish and loons and everything else that hunts them. They hold their position against the current, their narrow bodies pulsing, their tiny fins making constant, incremental adjustment, feeling out the river's flow. Each is perhaps one inch long. One shoots for a bug caught on the surface, shakes it, and slips back inside the group. They stand out against pieces of sunken drift; against the murk of the bottom they are almost lost. When I get down low and peer through the reflections I count maybe twenty or thirty. They will have been born, not quite here, but close to here.

Where to begin a story as cyclic as the salmon's? Had you been here in the autumn, in October, when the last of the leaves hang from the willow and the snows come any day, before the rivers are entombed, you would have seen little sign of them. But if you knew where to look, perhaps in that creek that came in by the trapper's cabin, there would have been patches of paler, upturned stones in the beds amongst the summer's growth of algae, maybe ten feet long by half as wide. These places are the redds (a Celtic word, 'to tidy up'), the Chinooks' spawning grounds. They were excavated by the female kings that reached here at the end of summer, lying on their flanks against the riverbed and shovelling with their tails, carving trenches well over a foot deep. Here they deposited their eggs, and the males that fight for space beside them ejaculated their milt, sometimes several males at once. Held together by their stickiness and the back eddy of the depression, the egg and seed fertilized.

As they start to develop, the eggs toughen and darken. The female works her tail again and begins to bury them, flicking gravel up into the current so that it falls across the bed. They settle into spaces deep amongst the stones. And as winter comes on, and the creek freezes over, the eggs begin to change. The speed of their development is directly governed by the temperature of the water. Within a few weeks two black specks peer out from the inside of each. Beneath the ice the current flows, scouring the rocks of silt and the wastes of growth and allowing the embryos to breathe. The temperature drops. At minus forty nothing moves except the raven. The air is so cold it cracks; too cold for it to snow. The smoke from the trapper's cabin rises straight up like a taut string. In midwinter, the salmon hatch. Huge goggle eyes on tiny heads, a translucent body with the line of their spine discernible inside, and the yolk-sac beneath their chin, like a goitre, attached by a single vein like the root of some bloody plant. Known as alevins now, they wedge themselves deep amongst the stones, turned by their instincts to face upstream so that the water flows across their nascent gills. The stars revolve; the sun grazes the horizon for an hour each day; the northern lights rush across the sky. The weeks pass and the yolk diminishes, and the alevins work their way up, day by day, through the gravel of the redd. Already they are imprinting the particular scent of this creek, the scent that they will use for their map home years hence. By the time they emerge from the creek-bed, over half will have already perished.

They hold there, in the tiny eddies of the stones of the bed, digesting what flows through their mouths, whatever microscopic organisms. They are fry. Each of their fins – dorsal, adipose, tail, anal, pelvic and pectoral – is identifiable now. They swim, they feed, on flagellates and rotifers and on tiny crustaceans, microscopic organisms as intricate as snowflakes. The days are warming; during the sunny hours the snows are melting. The ice above begins to crack, and one day, in a rush, it all goes out. A few will begin their journeys downstream now, but the majority remain for a year in these pools. The sun is high in the sky overhead and it comes unfiltered to the bed. As they grow they eat larvae and nymphs and other life that swirl about the water. They are silver beneath and speckled and olive on top and they are touched with the marks of the parr. And here we find them, this first morning, two thousand miles from the sea. The omen feels a good one.

We inflate the raft, and load the gear. Hector has chosen a raft for this first week because of the whitewater on the upper Nisutlin River. We have been unable to find much detail on the severity of the rapids except for Arthur Saint Cyr's map of 1897, commissioned by the Canadian government as part of a survey of 'all-Canadian' routes to the Klondike goldfields, on which Cyr has marked 'six miles of bad rapids' through the length of Nisutlin Gorge. A fish biologist who knows the area has advised us that they are rated Grade 6, which does not exist, is off the chart, almost biblical, and I can only hope that he has got confused with the number of miles. But that is still some days away. We paddle out into the current.

We are flowing markedly downhill, and the water, though flat, is swift; it would need a quick jog to keep up. The spruce here are vast. In this climate, at this altitude, their growth is glacial: some of the largest will be centuries old. It can be hard now to find decent-sized logs for cabins, all cut down years ago, but these would be ideal if you were so inclined to set up home here. Where they have fallen at the river's edge they lean out over the water, grasping; sweepers, they are called. The river moves fastest on its outer curve and often we are led straight to them; on the tight corners they are tricky to predict. Within ten minutes we have gone straight through one, and I have lost my hat. The raft is a pile of dead needles. In places these logs have detached, and run aground mid-river. Other logs jam on them, and then soil and silt build up behind. Eventually the willows take root, knitting the earth together, and islands form, islands we must dodge around.

By lunchtime the river has opened into Moss Lake, and the current fades away. A pair of Arctic terns, who see more light than any other creature on earth, hover, cruciform, far out above the water, scanning for young salmon. These terns have travelled the length of the planet to nest here, and in a few months they will fly back to the Antarctic. They have peaked caps pulled down low over their eyes; it gives them a debonair swagger. The sun glances off the water and there is no shade out here; I feel dizzy and sick with heat. We listen to the slosh of the paddle strokes. Ellipses and ovals and circles shimmer over the lake's still surface. Then in some hours we come to the end of the lake, and the current has us again.

With every few hours that we descend the river we race away from winter. There are spruce tips on the spruce now, a luminescent green against the dark of last year's growth. Spruce tips are sharp and lemony, almost like capers in a stew. A few hours more and the catkins on the willow are beginning to unfurl. We see the first aspen, and then later the first cottonwood, their new leaves so young and bright it is as though they have been varnished. And new flowers, shooting star, lupin, vetch, flecks of colour amongst the greens and browns, and then the first shoots of the fireweed, which Hector picks and adds to salads. We make camps on small islands and sandy beaches, and with the light there is little meaning between one day and the next, each are just pauses in the journey.

Three days from the lake, the McNeil pours into the Nisutlin through a canyon of rock, anticlines bent into its faces. The river takes the name of the Nisutlin now, a deference to the name of the Tlingit people, whose land this is, although the McNeil is much bigger than this tributary. Rounding a corner one afternoon we come upon a moose, standing alone on a gravel bar, sopping. It stiffens and sniffs the air when it sees us, but remains where it is, its stare fixed on us. There is a high-keyed hum of alertness. It is shocking, suddenly to stumble across another mammal out here. We stop paddling and float, the better not to scare it, watching it as we drift around the bar. Then all of a sudden Hector points and says 'Osprey!', and there it is, taking off from the top branches of a spruce with a fish held between its feet, and the moose bolts for the woods, picking its legs high like some huge improbable puppet, and it is all there in my field of vision, moose, osprey, fish, river, and it could be this time or it could be any other.

On the fourth day we come to the rapids. I had a sense of foreboding in camp as we rigged the raft that morning, strapping down the gear, attaching throw bags and flip lines, discussing eventualities: what to do if one of us falls out, what to do if both of us fall out, how we'd go about finding each other, what to do if we lost the boat. It is a long, long walk to Whitehorse. We have little idea what to expect. And then, for an hour or two we drift, along pleasant meanders, through sun-baked forests. We see goldeneye and canvasback, pintail and widgeon. Mountains recede in deepening shades. And I am almost lulled into forgetting what's approaching when we turn a bend and see ahead of us a line of rippling white, stretching from bank to bank.

'That's us,' says Hector, with a twinkle.

And we fall into them.

The mutter of the river becomes an all-consuming roar. The waves rise up, malevolent, great troughs and burbling peaks. The raft bucks over them, and we dig the paddles deep and try to hold her straight. Behind rocks that have tumbled from distant mountainsides are gaping holes that would flip the raft like a penny. The river wants to take the boat in there, and we strain to keep her out.

'Back up,' says Hector, unfalteringly calm. 'Sweep hard. *Harder. Now.*'

We swing sideways in a trench of water and the river sloshes over us, and shamefully I dive for the deck. The water barrels over me. The river heaves and flexes beneath my hands. The raft pivots where Hector jams the paddle in. And then we pop from the end of the rapid, soaked, unscathed, and I get back in my seat.

'That was meant to be the easy one,' he says, smiling, as though to say, well, we're committed now.

They blur one into another. We climb and plunge, the waves breaking in our faces, wiping the water away to see and bending back into the stroke. The river lunging, shaking, clawing at the raft.

'Well, look at that,' says Hector, pointing at ducks as we shoot over a wave. '*Another* pair of harlequin. How fascinating.'

I think that I am scared, but each time we emerge from the end of a rapid I find myself yearning for more. And there are more. They are continuous. We pick a line down the left bank, trying to dodge the worst of it, but the flow pushes us out into the middle channel, and we ride it, finding ourselves in an eddy on a sharp bend, the current pushing us one way, but not enough for us to miss the mound of rock that is now looming at the top of the next falls. I hear the edge in Hector's voice as we heave to try and pull right, and we do, grazing past the rock, gazing into the hole that yawns beyond it, and I'm looking behind, and Hector's shouting 'Back! Back!', and I realize we're rowing backwards past this next one, and we make it, just, turn to right ourselves, and on it goes. And all I can keep thinking is, the salmon swim *up* this.

And then, in a moment, it is over. The river broadens, slows, the land seems to flatten out. The sand bed sucks the current from the water. We lie back and float, and let the sun dry us off. I look across at Hector. We're both grinning, with adrenalin, with relief.

'I used to scare myself about once a week,' he says. 'Now I try and keep it to once a year. I hope that that was it.'

Some days later the Nisutlin meets the South Canol Road, the river's first scrape with civilization. Hector is to leave me here. He must return the raft, and Ann, a friend of Hector's who comes to pick him up, has brought me my canoe. There is just one small rapid left between here and the ocean, 1,926 miles away. The canoe is bright yellow, fibreglass, eighteen feet long, which is a big one, but for the many weeks of food I will have to carry further on down the river, when all roads and stores run out, that space will be essential. For now, it is a welcome change. There has been a fierce headwind, and rafts are not designed for that. The canoe holds its line better, it will be less of a slog. I can hear the car driving away for a long time before I am left there in the silence.

For four days I paddle down the slowing and widening Nisutlin. The sun is hot, the scent of warm spruce on the wind, and the landscape feels benevolent. I pass high-cut banks where the land has been bisected. Fifty, a hundred feet high, precipitous, topped with a line of spruce, the earth's inner workings exposed. There are trees which have slipped from the top and slid, and protrude now across the water. The hollows of bank swallows, *Riparia riparia*, pepper the exposed sand, neighbourly as blocks of flats, as though the cut bank had been used for target practice. Gusts of birds surround the boat, darting and dipping and chattering to the water. There are scarcely any insects yet, another symptom of the late spring, and I wonder what these birds will feed on, arriving at the end of epic journeys, or whether they have anything to feed on at all. A pair of mergansers in flight, necks flung out and quacking, hurry past me up the river.

It comes almost as a surprise to remember how alone I am. The Yukon watershed has a population of a quarter of a person per square mile for a third of a million square miles, the equivalent to a pre-agrarian society. If the planet were similarly populated the global population would be that of Istanbul, which is what it was seven thousand years ago. Out here, any evidence of human existence feels like an artefact, signs of other beings distant not through history but space. A tent peg, a shotgun shell, a penny: each object requires careful consideration, as much as a spoor print or a gnawed stick might. Once, on a bar where I have pitched for the night, foraging for wood, I come across the embers of another, older fire, scattered now to charcoal, and immediately I start, look up, as though aware, all of a sudden, that I am not the only inhabitant of this landscape.

One evening I stop to make a fire for coffee. It is ten o'clock and the sun is level with my eyes. Directly opposite a smudge of moon, like a thumbprint on the sky. There is a creek falling down between some banks, and I fill the kettle there and put it on the few sticks that are burning. I stand, stretching my shoulders out, waiting for the boil. The smoke of the fire rises thinly. There is a rustle of leaves up the bank. I start, jumpy for bears, and there is a fox high-stepping down the slope, through the tangle of fallen aspen. He hasn't seen me yet. He has an altogether different poise from the London foxes that I know, his coat glossy and thick, his tail bushy and high, like a rooster. He is several shades of red and rust and brown in different patches. He pauses, one forefoot raised, and drops his head, and then, assured that everything is as it should be, he carries on down the hill.

I lose sight of him somewhere in the scrub. I pour the coffee, piss on the fire, and get back in the boat, intending to drink as I float. Coffee to go, I say to myself, and laugh at my own joke, with no one around to tell me otherwise. I untie the bowline, and drift slowly along the bank, facing back the other way that I have come. The fox is there again, a little further down the bank. I fix the binoculars on him. He has a patch of white on each cheek, and he is looking right at me. I lower the binoculars, and look right back at him. I keep on drifting, backwards, gradually away. He keeps on watching, attention held. There is no rush. And he bends, and sticks his neck out, drinks, takes a final look at my diminishing form, and then he dances to the top of a log across the creek and carries on with the evening.

In a week I come to Teslin. The first village on the river, it sits on the west bank where Nisutlin Bay meets Teslin Lake, where the Alaska Highway crosses the water on a long bridge of steel girders. It is a three-hour drive from Whitehorse, but I have gone the long way round. *Teslin-to*, the thin long lake. About 450 people live here, the majority Tlingit First Nation. I moor up at the foot of a slope and climb it to the Yukon Motel

Restaurant. I order coffee and cherry pie. There are a lot of people, all of a sudden. There is British politics on the television. I am here to see Richard Dewhurst, a meeting organized long before I had got on the river. Teslin is the closest village to the headwaters of the Yukon, and so, as Richard tells me, 'we're the last people on the planet to utilize that fish', by which he means the Chinook. In his fifties, he is Game Guardian with the Land and Resources department of the Teslin Tlingit Council. He asks me if I would like to see his mum's old fish camp.

Richard drives a RAM 3500 Heavy Duty Cummins, all waist-high tyres and suspension and chrome. He is a large man but he looks small within it. He wants to know if I saw many moose on my way down the Nisutlin. I tell him I did, a lot.

'Bulls?' he asks.

I nod.

'Big ones? Big like that?'

He stretches his arms wide to demonstrate the size of rack, so wide that he almost spans the vehicle, and he smiles at the prospect. We bounce down a road off of the highway, a dirt track, really, overgrown with stems of alder. The Alaska Highway came through in 1942, part of the war effort, in case the Japanese turned up via Alaska. Teslin went, in a matter of weeks, from a remote bush village to an overnight drive to Vancouver. Elders remember bulldozers cresting the horizon and forging a path towards their village.

We come to a stop at the top of the beach on the lake, a few miles from the village, get out and slam the doors shut. In the silence, the water pushes softly against the pebbles of the shore. A ground squirrel runs along a pole of the collapsing fish camp, chattering at our disturbance. Once his family would have rowed up here. Richard walks through last year's leaves, running his hand over things. A couch with the foam poking out of it, the fake leather peeling back like a half-skinned animal. Floats strung up in the trees, a fallen flue, battered tubs for washing fish. The fallen corrugate of one structure has been placed sidewise around its wood supports, forming a windbreak three feet high, and two plastic chairs are set within it before a fire pit. Richard stares at it.

'That's new,' he says, after a time.

He rolls a cigarette from the pouch that he takes from the top pocket of his denim jacket. It is so quiet that I can hear it burning, crackling like twigs. He perches on the edge of a table that would have been used for the cutting of fish, one eye clenched against the rising smoke, and he peers back at me. A thin scar runs the length of his nose. This was his mother's fish camp, he says. They would come here in late July, when they saw the first wind on the water and the whitecaps further out, the whole family of siblings and cousins and nephews and nieces and friends, and they would stay here until late into August. Drying fish, smoking fish, picking berries, hunting moose. A place where the young could spend time with the old, where the kids heard stories, learnt to hunt, learnt to work as part of a team. The salmon blood brought in grayling, and the kids stood in the water with a line and a flashlight and a hook baited with salmon eggs and caught their first fish for themselves.

'They learnt how to run motors,' Richard says. 'And they learnt that water, eh. That water could take your life or it could save your life. You got to have respect for it. I do, anyways.'

He shows me photographs he has brought with him, tinted with age. Salmon laid out on the beach in a row, like a strike in ten-pin bowling. Himself, at camp, in his younger, trimmer days. His Dad, in brown shirt and Stetson, smoking a cigarette and squinting at the camera, holding up an absolutely enormous king, his mother grinning in the background. Another salmon, held up by his uncle.

'My uncle's got four or five inches on me,' says Richard, 'so you can see how big that fish is.'

The first salmon that they caught, they would welcome it, and thank it, for coming so far to feed them. They would roast it, guts and all, and everyone in camp would take a piece, and once there was nothing left but bones they would put the bones back in the water, pointing in the direction that the salmon had come, so that the next year they would bring more. When the salmon were running no one could swim. Back in the good old days they could put away eighty fish, enough to last the family through the winter. But that was the good old days. There was a time when Teslin was a ghost town during the summer, with everyone out at camp. He shakes his head, remembering.

'My mother's of an age now where she'll never see fishing here again,' he says.

Richard built the cutting room twenty-one years ago, and he got one year's use from it. The decline had started slowly, almost imperceptibly, although the elders had been warning them for decades that something was changing with their fish. Where Teslin is located, at the very endpoint of the run, they saw those changes first, both in the size and in the number of Chinook. To begin with they restricted their own fishing to five days of the week. Then it went down to three days. Then it was two. Nothing seemed to make a difference. Twenty years ago, all of the Teslin Tlingit Clans came together in an emergency meeting and they voted to shut it down entirely.

They agreed it was only temporary: just shut it down until things improved. But things had not improved. It has been tough, Richard says, to stand on the shore and see those salmon swimming past, tough when you know downriver in Alaska the people are still fishing. Tough when you watch those Alaskan TV shows and see them with their caches full of fish.

'Is it worth it?' I ask him.

'There are some people who think, if *they're* fishing, why are *we* trying so goddamn hard?' Richard says, grinding his cigarette between two fingers. 'But then you turn around and you look at your kid, and that gives you all the reason in the world.'

The salmon that comes to Teslin now is flown in from Atlin, another Tlingit village a hundred or so kilometres southwest, whose people catch their fish on the Taku River. Little of it is king, it is mostly sockeye and coho. It is expensive and it doesn't taste the same, and much of it is so rich and oily that people's bellies are not used to it. It dries differently from what they know, the ways passed down from their ancestors. And none of it comes with the heads on, which is the greatest delicacy of all. There are young people in the village who have never fished, who think that the drone of a bush plane from beyond Mount Bryde signals the start of the salmon run. Flying fish, they call it now.

Teslin did not fish for twenty years, but then, last year, they did. They made two fish camps, one on Teslin Lake, one down the Teslin River, and between them caught a little less than forty fish. Once they might have caught a thousand, but the idea was symbolic, ceremonial. There was a need to conserve the fish, but there was also a need to preserve the Tlingit culture. With those fish they cooked up a feast for the whole village.

'First time we ever set net for the generations so they can see our salmon when it comes up here,' Madeleine Jackson, one of Teslin's elders, explains to me the following day in the tribal government offices of the Teslin Tlingit Council. 'Man, we had five Clans here,' she says. 'I took one from each Clan to set that net, and when they came back next morning I took one from each Clan to run that net. So when they bring it out we had a ceremony and, man, there was tears of everybody, crying because first time they ever see that salmon. And some of those elders came and showed the kids what to do, how to cut fish, and how to respect it. It's not for us we're doing this. The elders had their share. I had my share. It's for the younger generations coming behind us.'

'The salmon are a giving people,' says Madeleine's nephew, Duane Aucoin, sitting beside her. 'They find their existence in a lot of ways by giving us life. And they give their lives for their children, when they spawn, in how far they swim. Their only goal is we need to do this for our children. What a good example for people to follow.'

He pauses.

'Now those Atlantic salmon,' he says, 'they're different. They don't die when they spawn. They swim out to the ocean and they tell their partners, see you again next year!' He smiles at me. 'That might be the European influence.'

In Richard's camp, I asked him how he enjoyed the feast. He shakes his head.

'I never went,' he says. 'I never participated. In my belief, there was one chance in a million that my son – he's five years old now – would ever get to fish. And I would never fish again if I thought that he could experience what I've experienced in my life. It makes me feel like we're cheating, that we're not being honest with ourselves.'

We stand there, side by side, looking out across the water. In the distance, in the shallows, outside one of the summerhouses by the highway, some chubby kids splash about on inflatable kayaks. Their shouts drift across the water. It looks as though, almost by consensus, every spruce tree has released its pollen on this morning, and there is a tide of its yellow dust along the shore. This old camp has the feel of a graveyard, of a place that has known grieving, but a place now calm, at rest. I listen to Richard's slow and heavy breathing.

'It sure was nice to hear those waves,' he says, 'when you lay down to sleep at night.'

Juvenile salmon in captivity display *Zugunruhe*, in the same way as caged migratory birds. That is to say, they become restless and more active at the same time as the other salmon of their species are migrating in the wild. The cones in their eyes alter to better prepare for ocean vision. They become more aggressive. In their tanks, they will orientate themselves to the corresponding compass points that their wild cousins are swimming to – to the west, to the north-west, to the south.

The parr set out in the first floods of their second spring, spurred on by changes in daylight hours and warming water temperatures, following the ice out to the ocean. To begin with they travel in small shoals, tails-first, so that they flow downriver, the same direction I am headed, ten miles a day, or twenty. As they travel their parr marks fade, their little bodies turning silver, in a process known as smolting. Other shoals flow in from other tributaries, and one day they will pass the adult kings that are bound the other way, upriver, swimming hard against the current, or snagged and thrashing in the set nets. In two months, or maybe three, the smolts will reach the delta. They will have eaten little, just what drifted into their mouths, but now they pause and feed. They eat plankton, sand-hoppers, shrimps, crustaceans, and they grow rapidly, adding an inch a month. Six inches, seven inches, eight. As the tides wash in they taste the first of the salt in which they will spend their adult lives.

Salmon did not always migrate. Whether they were originally a marine or a riverine fish has been much debated in the manner of the chicken and the egg. But now they are wired to travel. In *The Compleat Angler* (1653) the English writer Izaak Walton described the salmon as being 'like some persons of Honour and Riches, which have both their Winter and Summer houses', but the salmon's ability to survive in both fresh and salt water is rather more remarkable than that. Just half a per cent of thirty thousand fish species have the ability to straddle both these environments.

To be a porous animal in any aquatic environment is a challenge. In rivers, the water seeps into a fish's body, and its kidneys must continually expel it as urine, whilst at the same time cells in the gills extract salts and electrolytes from the surrounding water and transfer them to the blood, to prevent it from becoming too diluted. But as the salmon hit seawater, this physiology must change. In the ocean, water is drawn out of the body, and the salmon drink seawater to replenish it, their kidneys filtering out excess salt as the gills now reverse their function, excreting sodium and chloride. The kings will not know freshwater again until some four or five years later, in the last months of their lives.

I leave Teslin after three days. It is early June. Out on Teslin Lake it is hot, too hot to paddle, too hot to think. I float and drift and loaf. A loon is out there somewhere, warbling through its crazy cries. I sing dumb songs. I scare the ducks for something to do, like a small boy. Trucks rumble far off, out along the highway. All along the shore are the remains of other fish camps, old bits of metal glinting in the sun. Through binoculars, I watch two kids on a quad-bike scrounging one for firewood. I watch my paddle, the line and vortex of each stroke drifting away behind me like footprints across the water. I stop and swim and carry on, I stop and swim and camp. One evening I catch a grayling, and fry it up beside potatoes in my skillet on the fire. I have already forgotten darkness.

There are always those first few days, I find, until I shed the city, before I feel at ease again. Before muscles feel good, before cracked burnt skin stops hurting and feels like it's at home. Before my eyes open as wide as they ought. I dip a cup from the side of the boat and drink. Not even from a spring, straight from the lake. It feels astonishing that once all rivers would have run with drinking water, that once I could have dipped my cup into the Thames. And I remember the words of Bill Mason, the Canadian who did more to popularize modern canoeing than anyone else, who made it a rule not to paddle on water that he wouldn't also drink.

It is two days to the western end of Teslin Lake where, beneath a second bridge, it becomes the Teslin River. There is a diner here, and I stop for Eggs Benedict and coffee. As I get back on the water the weather shifts, I can pinpoint the very moment. The wind swings round to blow from out the north, gusting blackly across the water. The pressure drops out of the air like a stone, the clouds pile up, and then, for some days, it rains. The Teslin River moves quickly once it strikes out from the lake. The river's speed is shown in its reflections, a rushing lacework of mirrored clouds, ten miles an hour, twelve. The surface roils like cauldrons. Beavers rise beside the boat and, startled by my presence, slap their tails and duck beneath again. I slap my paddle back at them, the biggest beaver on the river.

The offshoots from the main stem are like a primer of Northern words: Log Cabin Slough, Muskrat Creek, Little Salmon River, Fish Hook Bend, Mosquito Gulch. Many other places on my map are named for the first white men who settled here, and speak to the diverse provenance of the prospectors who came into the country: McGregor Creek, Von Wilczek Creek, O'Brien's Slough, Johnson's Crossing, Erickson's Woodyard. I stop one night on the beach at Mason's Landing. In the woods back from the river, out of the rain, there are a cluster of collapsing cabins, built a century ago, following the discovery of gold on Livingstone Creek. Mason's Landing would have been the quickest way in: a float down the Yukon from Whitehorse to its confluence with the Teslin, poling the boat upriver to here, and then a tramp through the bush to Livingstone. Once there was a roadhouse and a stable here, a small trading post and a telegraph station. The police delivered the mail from Whitehorse twice a week. That is to say, in 1902, it was significantly easier to make contact with the outside world than it would be for me to do today. My quickest way to get a message out would be by paddling to Carmacks, about four days away. The buildings are fading back into the landscape now, overgrown with wild rose, sagging beams and fallen timbers. They look less man-made, more a peculiar constellation of natural elements. The roof of one is so thick with moss that it looks no different to the forest floor, and a thirty-foot-high spruce projects from it. On tin beaten into flues and back plates for the wood stoves, thick with rust, I can still make out the brand names of the products they once held.

All down the river are remnants from a time when the Yukon thronged with human life. Old gold-mining paraphernalia; the remains of roadhouses every twenty or so miles that were the stop-off points for travellers, rhubarb and raspberries still growing in their gardens; mooring rings drilled into rock. On Shipyard Island, the steamer *Evelyn* is rotting where she stands. Finding it amongst the trees is like coming upon some ancient Inca temple: 130 feet long, as tall as the spruce, with accommodation for eighty-five first-class passengers. Her hull splintered, the floorboards fallen through, the names of lovers scratched into her boiler. I walk along the upper decks, peering into cabins.

It is at Shipyard Island, two weeks into the journey, that I finally join what is, by common consensus, the main stem of the Yukon. Flowing from Whitehorse, it is swimming-pool blue, so clear one can see fish, but here the Teslin muddies it, and until it meets the ocean it will not run clear again. 'Yukon' is a contraction of the Gwich'in phrase *chųų gąįį han*, which translates as 'river of white water'. It is a milky, soupy brown. The silt, rubbed from distant mountains, whispers at the hull, and if you dip your paddle and hold your ear to the shaft you can hear it clearer still, as though the river is deflating. Each new tributary, many of which are big rivers in their own right, adds to the load of silt, so much so that by the time I have passed the eponymous White River it runs so murky that you cannot see deeper than a single knuckle beneath the surface.

The high-cut banks give way to basalt cliffs, their lower slopes thick with juniper and the vivid pinks of fireweed, now pushing out their first flowers. Fireweed is known as Summer's Hourglass here, for the creep of blossom up its stem can be read as a gauge to the proximity of winter, so that even now, in the endless light of early summer, there is the foreshadowing of its end. And the river widens, too; at points it is maybe half a mile from one bank to the other. Hoodoos, Gaudíesque, wind and weather carved, loom high above the river, hazed by ravens calling madly. One afternoon I see a wolverine, swimming crosswise to the current, like a piece of drift with a mind of its own. It climbs out of the river and shakes itself before it sees me, and darts into the scrub. A wolverine! When my grandparents moved into the house they live in now, they found a walk-in store cupboard covered with scratch marks on the inside, as though some beast had been kept in there. From then on it was known as the Wolverine Cupboard, although as a boy I heard it as 'wolvering', a verb. What sort of terrible creature could have wolvered this cupboard, I wondered. It has forever been an animal mythic in my imagination. And no less mythic for now having seen one.

Finally the weather clears and falls into a pattern of hot mornings, and a slow build-up of cumuli that become distant storms by the late afternoon. Off over mountains I watch the lightning, and one evening, perhaps an hour or so after a storm, a pall comes down over the river. The air smells of wood, like new planks on a hot day, and the river assumes an air of total stillness. Everything beyond the closest bluff is misted blue, receding lighter to the mountains, pale against an even paler sky. The smell of smoke comes stronger and catches in my throat. Then the day takes on a shade of sepia, shot through by the low sun, the exact same colour as the water, and I float through it as though through a dream. Later, in this murk, I hear music. I can see neither bank and cannot place where it is coming from. It is ethereal, far off, as though Sirens are calling to me. Eventually I convince myself that it is the engines of the planes that are bringing water for the fire. But still, as I drift through this dimensionless space, they seem to harmonize within me, and come from every place at once.

In a week I come to Dawson City, where the Klondike River joins the Yukon. The Klondike is a pretty river, and with the Yukon as my yardstick it seems narrow now and humble. It chatters over shallow rapids, and it borders many hundred square miles of tailings, the rubble left behind by the gold dredges, and the maze of pools they've pulled them from. Some of the tributaries that feed the Klondike run from the summit of King Solomon's Dome, and the names of those tributaries tell the story of this town. Eldorado Creek, Bonanza Creek, Gold Bottom Creek, Last Chance Creek, Independence Creek, Little Gem Gulch, Nugget Gulch, American Gulch, Oro Grande Gulch, Too Much Gold Creek, All Gold Creek, Not Much Gold Gulch. 'Klondike' is an English fudge of the Hän word *Tr'ondëk*, meaning hammerstone, a hard cobble used for the hammering of stakes into the riverbed for the setting of the salmon weirs that would guide fish into the traps. Klondike, an onomatopoeic word, the sound of rock on wood, and the rush of water over rock, and a word that screamed from headlines and spread rumours in the drinking holes, a word that seduced a world in the depths of depression. I float into Dawson City in the same way that a hundred thousand other people would have turned up, a little more than a century ago.

These days Front Street has been abandoned to the tourists. It has kept its wonky boardwalks, and left its streets unpaved. Diamond Tooth Gertie's Gambling Hall has three cancan shows a night. The poet Robert Service's cabin is here, and one half of Jack London's (the other half a museum in Oakland, California), two men responsible, perhaps more than any others, for writing the fictions of this place. By law any new buildings must give the impression that they were constructed at the end of the nineteenth century. In Maximilian's Gold Rush Emporium you can buy nuggets and antique panning equipment and mammoth ivory and Jack London first editions. Men and women dressed in petticoats and buckskin will show you around the historic post office, the historic bank, the historic paddle steamer; they will tell you stories of courageous men and lascivious dancing girls, and strange things done under the midnight sun. Every night in the Downtown Hotel, its original front wood-panelled, its wooden bar running the length of the room, there is Dwayne, all gums, trousers hitched up high with braces, wasted, hammering through his repertoire on the upright piano, ragtime versions of 'William Tell', 'Drunken Sailor', 'Head and Shoulders, Knees and Toes'. He sips from a water bottle. Later, outside, I take a sip; it is vile even for moonshine. 'Made from instant coffee,' he grins at me, toothlessly. At the table opposite, a man in a beard and captain's hat serves up Sourtoe Cocktails to the passing tourist trade, a shot of liquor with a human toe afloat in it, like a blackened monkey nut with a nail. *You can drink it fast, you can drink it slow*, state the rules on the wall, *but your lips must touch the toe*. The implication is that in a frontier town, far from centres of rule and regulation, this is the sort of thing that the residents would get up to, that we are only a few laws away from drinking body parts. Swallowing the toe is expressly forbidden, but it has happened on occasion. The fine for swallowing was five hundred dollars until one evening a man deliberately necked it, slapped down the money and walked out. The fine has since been raised. This is toe number ten. People donate toes in their wills. Once, hitching south out of Dawson, I was picked up by the town's only resident cop, whose work that week was taken up in tracking down the toe, which had been stolen the weekend before. During the ride he received a call on the radio from a man purporting to be the thief, and he made me step out of the vehicle while he interrogated him. A charge for trafficking human remains is as serious as you might think. The toe was later returned, by post.

Dawson is a mix both of couples touring and of those who have blown in hitchhiking, in search of work tree planting or morel picking or something else cash-in-hand and back-breaking. The tree-planters live out in the woods, dodging the twelve-dollar fee at the government campground, a few people and a few dogs to a camper, hustling for work and eating out of dumpsters. The couples inhabit the RV campgrounds in immense and gleaming vehicles with names like Prowler, Hitch-hiker, Bounty Hunter, Wild Cat, Ultimate Advantage. Many will have sold their homes upon retirement and moved into an RV of equivalent value, migrating south to Texas for the winters. Snowbirds, they are called. The two tribes prop up the bars in different shifts, swapping stories of the road. The bars do not shut. People mill in the dusky streets at 2 a.m.

'What's up?' a man calls across the way.

'The price of gas,' shouts back his friend. 'My spirit. Maybe my dick if you had some tits.'

Dawson City is a six-hour drive from Whitehorse, four hours

across the border to the first town in Alaska (Chicken, population twenty-three, winter population seven): it is about as frontier as it still gets. In one of the tourist bazaars the motto of the Yukon Order of Pioneers is hammered to a crossbeam: 'Do unto others as you would have others do unto you', and it is this anarchic ethic, imported to the North in the 1890s, that still pulls in the crowds today. The code of frontiersmen running from themselves or something else, who rejected hierarchy and class, and who relied on each other implicitly because there was nothing else upon which to rely. The get rich quick spirit, too, remains; the tough guys can still strike it big here. You see them in town, the Yukoner style. The grey hair, not long, but down to the shirt collar; not anti-authoritarian long, but the look of a military man who's thrown in the towel, squirting out above the ears from beneath a baseball cap, the cap emblazoned 'Canada' with a stitching of the flag. A clump of moustache above their lip, like a draught excluder. The check plaid shirt, probably tucked into jeans with a belt with a brass buckle, stretched over a slight, tight belly, and a pair of leather work boots that you hear coming on the boardwalks. They sit in loose groups in the cafés on Main Street, drinking thin coffee and eating pie, discussing pistons and bears and the weather, keeping an eye on the river and the pretty girls that pass. Elderly men emerge from their RVs, white-kneed, and try their hand at gold panning on the tours: this could have been me, they think, a century ago, if only the wife would have let me. I would have made a killing.

The Klondike's first big strike came in 1896. In 1897 the population of Dawson hit 5,000. By 1898 it was 40,000. By 1902, the city had government buildings, a power plant, four newspapers, several churches, a library, a court of law, a water-works department, several schools, a swimming pool, a bowling alley and a curling club. But gold extraction had already peaked. In 1900 a million ounces were taken from the ground; in 1904 it was down to 400,000 ounces, and the population was down to 5,000. Today the population of Dawson City is 1,375. The whole town is riddled with nostalgia for some confused and distant time when brave men wrung fortunes from the lands like water from a dishcloth, lands that Jack London, who spent fewer than nine months here, and most of those in bed with scurvy, described as 'new and naked'.

I walk into the Visitor Centre. Its centrepiece is two interlocked racks of antlers, moose that had locked in battle and later died when unable to separate. *Fight to the Death*, the piece is called. It is hung with flags and red and white streamers to celebrate upcoming Canada Day. There is a man looking at it in puzzlement.

'I don't think this has really been thought through,' he says.

Dawson is a palimpsest, another story sits behind it. The missionary Hudson Stuck, writing in 1917, was a hundred years, if not more, ahead of his time, when he said that 'the great stampede to the Klondike of 1897 and 1898 brought nothing but harm to the native people'. Do unto others as you would have others do unto you.

A few blocks back from Front Street, I visit with Percy Henry, a Tr'ondëk Hwëch'in elder. There are few First Nations people on the tourist streets, but on the blocks behind the strip it is a different, quieter town. Percy gestures to where coffee is brewing on the counter, and I pour myself a mug marked with a verse out of Galatians. There is a vibrant, psychedelic painting of Jesus in a frame, and tacked to the door a printout reads: *TOGETHER We can all be the stewards of God's creation SO COME ON . . . let's save the planet*. Percy sits at the table forking at a microwave dinner of meat and swede and peas; the skin on the backs of his hands is as thin as tissue paper. He is eighty-nine years old, and he has the wild and playful eyes that people get when they realize they have hit an age where they can get away with anything.

If ever there was an advertisement for the health benefits of wild fish, of clean water and clean air, it is the elders that I meet here. In 2000, Percy's parents were entered into the *Guinness Book of Records* as the world's longest married couple, eighty-two years long. Joe Henry quit fishing at 92, and died at 104. Annie died at 101.

'So,' says Percy, putting in his hearing aids. 'What's your problem?'

Two years ago, the Tr'ondëk Hwëch'in took the decision, following Teslin's example, to stop fishing king salmon for a seven-year cycle. It was a voluntary ban, and it has mostly been respected. The river has been quiet. The youth are taught to cut fish with chum salmon from the freezer. It can be hard to generate the same excitement. I ask Percy what he makes of it all.

'King salmon can send message to others if they in trouble,' he says. 'I work out that and I try to tell Fish and Game. They wouldn't listen. They fool around with fish too much, give them tag and number. I don't know why today they monkey around with fish. King salmon is our food. Sure all the fish are edible. But king salmon is something, eh?'

He sits back, and I wait. I have learnt not to interrupt when an elder is speaking.

'Last fall we caught sockeye here,' he says. 'That's Fraser River fish. In Alberta they're squeezing oil out of the sand. Now they poisoned the big river. The sea is dying. When you see those big whales come shore to die you know there's trouble. They got no place to go. I shouldn't just say white people, but they cause a lot of trouble. They know it's going to happen, but the dollar comes before. The fish didn't change. We change.'

Before 1897, life had not changed significantly for the Tr'ondëk Hwëch'in for several thousand years. Each year was cyclic, as families moved through a series of camps, following the resources that they lived by. By the time that the last of the fall chum were dried, the summer was nearly at its end. In the final days of warmth they picked and preserved berries and roots, and built the snowshoes and sleds that would see them through the winter. As the autumn drew on they moved up into the hills, hunting caribou and sheep, drying the meat until it froze and storing it in caches that they would return to later. In winter they dressed in furs and lived in rapidly constructed dwellings insulated with moss that they lined with caribou skin. Late in the winter they hunted moose, and by spring they would be at the head of the Klondike, high up in the hills, trapping beaver and muskrat and birds. When the ice went out they built boats out of mooseskin and floated down to the Yukon. This was the time of gatherings, and sometimes, as a boy, Percy remembers his family poling their canoes upriver to the White River, where they would meet Tutchone and Tanana people. It was a time to see friends and share food, for couples to meet and marry, and then they would float back to the Klondike to be there for when the kings came. The details changed, of course: where they went, and who with, and for how long. But there was constancy and rhythm.

If this was a typical summer, Percy and his family would already be out at fish camp, getting ready. They have already heard the first of the thunder, the salmon tails hitting the water in the distance, saying we're coming! Prepare yourselves! Percy remembered the enormous fish of his childhood. They worked all day, cutting fish and drying fish, and when it was time to eat, people would sit down in a circle and put in the middle what they had, bannock and blueberries and dried meat and fish. If you were too old to fish, or too poor, or too weak, then people would look after you. They would take as much fish as they needed for the year and let the rest carry on upstream.

Percy smiles at the memories. 'I'm not real elder,' he says, leaning forward in his chair, conspiratorially. 'I'm old, but I experienced modern day. If you listen to *real* old people, before white man, they can tell you *lotta* things. *They* can tell you future.'

Joe Henry had lived to see the future. Percy's father was born in 1896, the year of the first gold strike, and he died a few years after a ban was placed on the commercial fishing of king salmon by bureaucrats he had never met, living in a city he had never seen. White men had been turning up here and there since the 1870s, gold prospectors and fur traders and missionaries and government officials, but it was with the discovery of the Klondike gold that the floodgates really opened.

It was late spring, 1897, when the first influx of goldrushers arrived, and within weeks the Tr'ondëk Hwëch'in had been displaced from their fish camps on the Klondike. The natives crossed to Dawson, only to find that the land had already been occupied by speculators and the Royal Canadian Mounted Police. In the space of one season they were evicted from lands that their people had known for ever. Their traditional fishing grounds had showed scant trace of their millennia-old presence except for some stakes to set the weirs and a few simple dwellings. Now there were several hundred tents, a sawmill, two saloons, a red-light district and a typhoid epidemic. The Tr'ondëk Hwëch'in called it, in the new language they were acquiring, Lousetown.

An Anglican missionary, Frederick Fairweather Flewelling, purchased forty acres at Moosehide, two miles downriver on the Dawson side. He built a church, and gave the rest of the land over to the displaced natives, yet despite this gesture, Flewelling could not hide his contempt: 'They have as a race neither worldly nor spiritual ambition . . . They have few or no traditions and have as a race no individuality . . . Work among them is difficult because of their nature and because for nearly nine months of the year they are off in small bands hunting and fishing.'

Hunter-gatherers are perceived to be nomadic, but it is quite the other way around. Certainly the Tr'ondëk Hwëch'in and other First Nations were in constant motion before the white man arrived and built for them their churches and their schools. But they remained upon their land. It is the pastoralists, the farmers, who are the restless ones. It is Columbus and Cortés, urbanites both, who came to the Americas. There is a distinction to be made between nomadism and restlessness, and it is restlessness that drives one further, beyond the lands one knows. It was city-dwellers who went to the moon.

Percy is one of two surviving speakers of the Hän language, one of the seven languages of the Tr'ondëk Hwëch'in people. He cannot read or write. He has a computer, which he uses for listening to an audiobook of the Bible, in the same way as his mother used to read to him from a Bible written in Hän when he was a boy. Percy made it through school as far as the end of Grade 2 when one of the other pupils punched him in the head and knocked him out. When he came round he was perfectly unharmed, except for having forgotten his entire formal education. He told his father he was going to quit school.

'What you gonna do then?' his father said.

'I'm going to go to *your* school,' Percy said, and that winter he started with his father on the trapline. He still believes that punch to be the best thing that ever happened to him.

At age twelve he got a job as a fisherman's assistant, working eighteen-hour days for a dollar checking nets on the river. Since then, in the frontier spirit of anyone who makes a go of it here, he has done almost every job going. He was nine years in the sawmill in Mayo. He worked in logging and for the highways. He has been on fire crews and has captained the Dawson ferry. For several years he worked on the barge that hauled supplies to the village of Old Crow, on the Porcupine River.

'I first went to Old Crow 1943,' he says. 'I met *lotta* elder.

Lotta elder alive then. And they tell me story. They always tell me story. The ducks there in Old Crow, there are *million*. They tell me you see that? I say yeah. If you live long enough to see what we're going to tell you, in the future there'll be nothing.'

Percy looks out of the window. 'This highway, you used to drive through ducks. Ducks! Now you don't even see *one*.'

Recently in Old Crow, his friends have told him, a lot of the female caribou have stopped calving.

'That's a bad news,' he says. 'They're finding a lot of caribou dying, because the doctor quit bothering with them.'

'Who's the doctor?' I say.

'The wolf is the doctor of all animals,' he says. 'He chase caribou. He don't kill 'em right there. He could. But his mother train him, you don't kill 'em till one falls side. That's a weak one. So that's how they stay healthy. Make 'em sweat. You see that Yellowstone Park. Animal there were half dead. So they took some wolf in there and all the animal were happy. Bring their life back to where it should be.'

But now, Percy says, the young wolves don't know what to do. It began when the state started culling wolves as a way of protecting caribou. They shot the old wolves, the ones that train the pups. Now Percy sees wolves coming into yards to attack dogs, he sees wolves chasing skidoos. They haven't been taught fear; they've had no education from their elders.

'But people don't believe me,' he says, without resentment. 'Because you can't read it in a book.'

In 1968 Percy was elected chief of the Tr'ondëk Hwëch'in, a position that he held until 1984. I walk around the room, looking at framed photographs on the walls of Percy with various Canadian prime ministers: Jean Chrétien, Pierre Trudeau, Justin Trudeau.

'That time land claim start too, eh,' he says. 'Non stop. I was so busy. First of my kids I hardly know them, they don't know me.'

Land claims gained momentum in Canada through the 1970s and 1980s, as more and more First Nations challenged the Canadian government's occupation of lands that were ancestrally theirs.

'I was so busy that finally the wife left me,' he says. Mabel came back though. They have been married a mere fifty-eight years. 'Sometime we go down,' he says. 'Then we go back up.'

Percy has recently had a new outboard motor delivered, a Yamaha 150 horsepower. It is on the table in the front yard; he would like it to be in the shed. He had a heart attack a few years ago, back during his drinking days, and since then he is forbidden heavy lifting. He finds it hard to get anyone to help out. But now I've shown up through the door.

It is awesomely heavy. We grunt and heave. Yet after twenty minutes of rigging ramps and slings with bungee cords and lumber, we admit that there is no way we can move it between the two of us. I collar a middle-aged Australian with thick tattooed arms in the RV park across the street. Together the three of us manhandle the thing inside. It is a curious tableau. Afterwards we all shake hands and the man goes back to his deckchair. I lean against the shed, getting my breath back.

'The world is good, but we ruined it,' says Percy, continuing an earlier train of thought. 'They got millions and billions of car. The government, the world, too busy trying to make better plane, better bomb, to kill more people eh.'

I ask him what he thinks we can do about it.

'We can't stop it now, too far. We all up shit creek.' He smiles at his own swearing. 'I think our kids will starve. I don't know if you believe in Bible, but everything it say there is happening. Right now. I study Bible. I study the elders' stories too. They're not too far apart. In Bible, God say I will never do it again. Well, this time we going to do it to ourselves. We're going to destroy our home.'

He chuckles at the madness of it all. 'The other animals live by God's commandments. What can we do? Die maybe. That could be our cure.'

The old Hän fisheries factory on Front Street, built of corrugate, looks out over the river. It has been shuttered now for years. This is one historic building that the tours do not take in. But if you wanted to look for the first clues to the king salmon's collapse, for where the end began, then this building would be one place to start.

Next door to the factory, inside the Tr'ondëk Hwëch'in fisheries department, tacked onto the side of a filing cabinet, is a black-and-white photo, scribbled with a note: 'King Salmon (85lbs) Caught in Dawson, YT, July 26th 1924.' A man looking spiffy in a three-piece suit, flat cap and tie, on the boardwalk outside Jimmy's Place, is staring at the camera. Beside him is his fish. It reaches from his cap to the bottom of his shin. June hogs was one of the names they were once known by. It is monstrous, a great reptilian torso. The biggest king ever caught on a line was 97 pounds, 4 ounces. In 1949, near Petersburg, Alaska, a 126-pounder was caught in a fish trap, the upper limit of a featherweight boxer.

If the salmon is the King of Fish, as it is so often called, then the Chinook is the King of Kings. Their scales are a burnished chrome, and each individual scale is picked out by the light so that it stands out, precisely, from the rest. In the ocean this makes for perfect camouflage, shimmering against the water's light. They fade from the pale white of their bellies to a dark grey along their backs, so that predators above cannot distinguish them against the ocean bed, and those swimming beneath cannot tell them from the sky. Their backs, and their fins, are speckled, to enhance further the effect. A single line of darker scales, like a horizon line, runs from nose to tail. In Atlantic salmon this line is magnetized, in the manner of a compass, to help them find their way (the Pacifics carry their magnetite in their skulls). They are broad and stocky creatures, and they appear full of muscle, inflated with it. The Chinook's flesh ranges from pearl white through pale pink to the deepest, richest red.

The first time the flesh of the Yukon River king was given monetary value was when it was sold in camps during the Gold Rush. Many men had arrived woefully unprepared and many were starving to death. In 1914 an agent for the US Bureau of Fisheries made a trip from St Michael, near the Yukon's mouth, to Whitehorse, and concluded that commercial fishing on the river would not be profitable or advisable. He noted the crucial role that salmon played in maintaining the river's ecosystem: for natives, for white people, for dog teams. Hunting was, by its nature, fickle, but the salmon runs ensured a glut of protein that arrived on schedule, year on year, and made life in this climate possible. The subsequent report warned that 'there is strong prejudice against the establishment of canneries on the lower Yukon, if such an undertaking should ever be considered feasible, as it would mean cutting off or greatly reducing the supply of salmon up the river, the result of which would be great privation and hardship to the people of that district'.

The first cannery, owned by the Carlisle Packing Company, came just four years later. Along the Alaskan coast, there were already canneries supplying half the world's salmon supply, but the first on the Yukon met with resistance. Chief Paul of Koyukuk was concerned that it would take 'every fish that came up the Yukon'. In 1917 a poor run of salmon had led to villagers being forced to kill sled dogs because they had nothing to feed them on. Dogs were as crucial to river life as horses were elsewhere: they were needed for hunting, for hauling and for travel, and for delivering the mail. Locals were reluctant to see anything that might further impact the salmons' run. The Carlisle company replied that if the natives didn't put up enough fish, they were lazy and their fishing methods primitive. In 1919 the cannery caught over a hundred thousand kings, a number that would not be bettered until 1961.

It is true that runs have always fluctuated wildly. Gluts were followed by years of scarcity, when high water and poor weather could make for a year of hardship. In 1919, the year after Carlisle opened, many people went hungry, and one local priest testified that 'the wolf really is at the door'. The US Bureau of Fisheries reminded natives that there were plenty of other types of fish, not to mention all the game, and besides, no report of undue privation had been substantiated. 'Consideration must undoubtedly be given to the psychological effects of the establishment of the cannery on the Natives; they heard the cannery was in operation, hence at once assumed that there would be no salmon passing to upper waters,' they wrote. The presiding judge tasked with investigating the matter spoke of 'the rights of private property in a free society and the sacredness of honest investments made in good faith'. Two ways of envisaging the world were being forced up against each other.

Yet despite the Bureau's confidence, they sent Dr Gilbert and Henry O'Malley to investigate. They found that the run of 1919 had been 'one of the worst, if not the very worst ever known on the Yukon', and that had the hunters and trappers not been particularly successful that year, the winter would have been catastrophic. Just how much Carlisle's fishing had an impact on the run was hard to ascertain, but both local interviews and their own assessment indicated that the cannery was exacerbating what was already a bad year. Gilbert and O'Malley suggested that it was wrong 'to experiment with the welfare of the people of the Interior'. In 1924 commercial fishing for export was banned entirely. Carlisle packed up and left for Bristol Bay, south-west Alaska.

But in the 1930s aeroplanes began to replace dog teams, and with salmon no longer needed for dogfood, commercial fishing reopened, albeit at its inception with more rigorous restrictions. When Inspector C. F. Townsend, who had inspected the fisheries for more than twenty years, retired in 1942, he wrote of how industry was bringing a much needed income to 'the few whites and all the natives in this vicinity', but that 'I would never recommend any increases in the limited catch now in force for just as soon as the limit is increased I am afraid trouble will begin.'

This balance held for a time, but the 1960s brought powerboats and monofilament nets, and the improved technology enabled a new fishing style of drift netting. A three-hundred-foot net, dragged through the middle channel of the river, could harvest kings at levels and in places that had been impossible before. In 1960 there were forty-six drift nets on the Yukon; by 1975 there were 314. In 1976 the Alaska Department of Fish and Game (ADF&G) released 700 drift-netting permits. Harvests boomed. In 1980, 150,000 kings were caught commercially, a number that has never been bettered. ADF&G interpreted these numbers as a sign that runs were improving, and upped the quotas accordingly.

The Hän Fishery plant opened in Dawson City in 1981, and Minister Don Meeks started running a riverboat to buy up catches and take them to the plant. Then they put a road in at Fortymile and all the fishermen would gather there at eleven each morning to help load up the catch. For a time business boomed, and money was good. Some days they hauled in 14,000 pounds of fish. 'A huge human effort,' one man tells me, with the glee that comes from the nostalgia of hard work, in the years before he got a desk job and a belly. 'It wouldn't be accurate to say I depended on the king, but it sure made life a lot easier.'

But the price fell out of the market as farmed fish began to dominate. And concurrently, the numbers and sizes of wild salmon began to collapse. In the 1980s the buyers of Chinook would refuse to take anything smaller than a fourteen-pound fish. Then they dropped to twelve pounds, then to ten. In the few years preceding 2007, when the Canadian Yukon's commercial fishery was shut down, they would buy anything they could. In 2006, Dan Bergstrom, ADF&G's management supervisor for the Yukon River, said: 'We don't see a crisis at this point.' State and federal agencies on both sides of the border declared either 'economic disasters' or 'fishery disasters' in 1997, 1998, 2000, 2001, 2002, 2009, 2010, 2011 and 2012, and yet commercial fishing was not shut down on the American side of the border until 2011. By that point, the bottom had fallen out of the run. The historic average, before 1997, had been 300,000 fish. In 2013, 37,000 fish came back. In 2014 a complete ban was placed on all fishing for kings, subsistence included, on both sides of the border. Never before had the subsistence fishery been banned, and it left a lot of people hungry, and angry.

'We were raised to be on that river,' one Dawson elder, Peggy Kormendy, told me. 'It was the *strangest* feeling when they said you couldn't fish.'

What made it worse was that no one was able, or willing, to explain why it was happening. Fishermen stood on the banks all summer and watched the salmon pass.

In 2011, Debbie Nagano was entrapped by two fishery officers posing as tourists from Alberta, and fined $5,000 for selling Chinook.

'The court case was totally crazy,' she tells me when we meet. 'They pointed the finger at me and they said "you should have known better". After I went through it, I didn't even want to look at a fish. I was too angry. I still can't even talk about it. If they came to me and said "Are you selling fish?" I'd have said "Yes, I'm selling fish. You know I'm selling fish, I been selling fish all my life." You gotta keep yourself busy, that's why we fish. And if they take that away from you, holy crowly. Then what you gonna do?'

In 2011 limited trade was still permitted between First Nations. I ask Debbie why she didn't just sell to her own people?

She shakes her head. 'They come to you and they say "Come on, we've not had it for a long time." I wasn't raised to say you're non-native, I can't sell to you. When the white people first turned up here, a lot of them had nothing. No meat. They save a lot of non-native people, these native people.'

Debbie despairs. To hear the stories that her elders tell her, of how harmonious this place once was, and to see the pollution from the gold fields and the asbestos mines leeching into the rivers, and the destruction caused by the tourist boats, and how poorly all of it was managed by the managers for so long. To be pushed around by laws that once there were no need for, to be governed by people who have no respect for the environment. And now they want to frack up on the Dempster. That's God's country, her grandpa always told her.

'We want our children to see what we saw in our lifetimes,' Debbie says. 'And if we continue to be disrespectful, they won't know.'

The second night out of Dawson, I make camp on flat ground in the middle of the river. I have been out for several weeks now, and the daily chores are feeling less like chores, and more like necessities of life that it makes no sense to question. They are mundane, in one sense – the couple of hours it takes each evening to unload the boat, to raise and stake the tent, to gather wood, to make a fire, to cook dinner, to haul the canoe up the beach and flip it and tie it down to something solid in case the wind picks up in the night, and then two hours more in the morning to reverse the entire process – but they are no more mundane than a tern finds it mundane to build its nest, or a beaver, working at a willow. There are things that my life here dictates I must do, and so I get on with them.

My awareness is changing. It is becoming increasingly easy to read shapes into the landscape, to see a pair of moose antlers in a distant piece of driftwood, or a man, waiting for me, in a stunted birch: with my mind left to wander I am inventing curiosities, and I imagine that it will not be long before I start weaving them into stories. I am noticing irregularities, too, so that I am able to focus in on a fleck of white from half a mile away, and spot a bald eagle sitting motionless, scarcely aware how I have done it. I find that I can tell a species of a tree by how it is moving in the wind, how the aspen leaves twinkle but the birch's quiver. I have never noticed this before. It is the same with birdsong. I had always thought that learning birdsong was beyond my capabilities, but out here the songs are starting to stick: the dark-eyed junco, which sounds like a telephone ringing; the white-crowned sparrow; the raucous kingfisher. Despite my many years of city living, I think perhaps I might not be a lost cause after all.

I am also falling in love with the canoe. It has an elegant simplicity, the essence of boat: symmetrical on both planes, with only the position of the thwarts to indicate which end is fore and which aft. The longer I spend in it, the more I realize that I can feel the river through it, that it is more a conduit between me and the water than a vehicle separating one from the other. Each flex of current, each stir of wind, can be felt through the hull and through the wooden paddle, and the more I pay attention, the more I feel. Bill Mason, the legendary Canadian canoeist, again: 'When you look at the face of Canada and study the geography carefully, you come away with the feeling that God could have designed the canoe first and then set about to conceive a land in which it could flourish.'

The original designs were built of birch bark. A frame of white cedar, covered with sheets of bark peeled from a birch tree, lashed together with spruce roots or caribou tendon, and waterproofed with spruce gum. They weighed perhaps ten pounds. The first white travellers wrote with astonishment of the skill with which the local people controlled them, their paddler standing to spear a salmon whilst the canoe held fast in whitewater. The beauty of the design meant that they could be repaired with surrounding materials, and when a canoe was beyond redemption it could rot back into the forest. My canoe, on the other hand, is made of fibreglass, and it will last until the end of time. But what is sacrificed in sustainability is made up for in lack of required skill: this canoe will bounce and scrape off almost anything. Down on the coast the Tlingit built their war canoes from the logs of giant cedars, and they could seat up to forty men, hollowed out and steamed open by placing hot rocks in the dugout. They still make them in Teslin this way, for the annual summer races. And whilst everyone living here now uses skiffs, the simple, aluminium boats, and outboards, for daily life – I will not see anyone native in a canoe for the entire trip, and indeed, the younger generation are often nervous of them – there is still a deep-seated cultural curiosity in the way that I am travelling, the same as I might reserve for a Tlingit man who decided to travel through London on horseback for his holidays.

I sit up late. Blocks of ice lie around like walrus, stubborn in the sunshine, grubby with the endless passing of the silt. I listen to them drip. Feet deep, they are the herald of another season: this was the river's carapace before the spring broke through. Onshore, the spruce are scoured by ice, their bark ripped to shreds, like the scratching posts of some great beast. A thrush sings, like our blackbird, calling in some meagre dusk. Twilight is so drawn out here that astronomers separate it into different twilights – civilian, nautical, and astronomical – depending on the number of degrees that the sun has dropped below the horizon line. Tomorrow, I will be in Alaska.

Alaska is delineated by Bering's straits. Alaska, from the ancient Aleut *alaxsxaq*, 'the object toward which the action of the sea is directed' or, in another translation, 'the shore where the sea breaks its back'. Alaska was as much a draw for Russians expanding east as it was for Europeans expanding west, but they came for fur, not gold. In 1728, Vitus Bering, the Danish explorer, sailed between Russia and North America captaining a Russian expedition sponsored by Peter the Great to determine whether the two continents were connected, and the strait he discovered that held these two land masses at arm's length, fifty-three miles at the narrowest point, would come to take his name. In 1741, on a subsequent voyage, Bering sighted the conical, volcanic peak of Mount St Elias, from which, in 1825, a line would be drawn straight north up the 141st meridian, as far as the Beaufort Sea, demarcating British Canada from Russian Alaska. Bering's crew returned from that voyage with eight hundred sea-otter skins and sold them for a fortune. The first of Alaska's rushes was on.

Promyshlenniki, they were called. The fur traders. Far from home, unencumbered by governance or morality, they moved east across the sweep of the Aleutian islands, from one island to the next, decimating populations of animals and people. They took the women and children hostage to ensure the male hunters would bring them their pelts. Occasionally the Aleuts murdered their new masters; they had never encountered hierarchy before. The Russians settled villages and established posts along the lower Yukon. That Russian heritage is preserved in the surnames of those that live along the Yukon, names like Demoski and Kozevnikoff and Shaishnikoff, and in the onion domes of the Orthodox churches in the villages downriver.

Today, Mount St Elias is the second tallest mountain in both the United States and Canada. The bush is cleared ten feet either side of the boundary line, a scratch mark that splits the countries north to south. Every twenty-five years chainsaw crews head out and cut everything back to the ground. From the river it is blink-and-you-miss-it, a single crenellation in several hundred miles of unbroken forest. For all the brutal anonymity of most international borders, this one feels charmingly homespun. There are far more animals than people who cross back and forth across the line. This is where the salmon switch their nomenclature from the Canadian Chinook to the American king.

On the Alaskan side there is a fish-counting sonar, which will monitor and record every salmon that comes upriver, although for now it is still boarded up, no work to do just yet. (The leading Chinook scouts are now five hundred miles away.) But as a human, if you wanted to sneak across the border undetected, this might well be the place to do it. In the village of Eagle, eight miles into Alaska, there is a phone in a cabin with a direct line to US customs. I pick up the receiver to let them know that I am here.

'Do you have a shotgun?' says a voice on the other end.

I tell him I do not.

'Very good,' he says. 'Enjoy your visit.'

I walk around town. There is a hotel that appears, from its size, to be expecting far more people than will probably ever arrive. I sit in one corner of its echoing restaurant, drinking refills of coffee. I washed specially in the river this morning, but inside I feel quite dirty. On a pole, beneath the Stars and Stripes, snaps the Alaska state flag, designed in 1927 by thirteen-year-old Bennie Benson, an Alutiiq boy, as the winner of a statewide competition. Against a dark blue background are eight yellow stars, spread out to represent the constellation of the Big Dipper and Polaris (the North Star), but the overall effect is more of an EU flag falling to bits. In the City of Eagle Amundsen Memorial Park, just off Amundsen Street, is a globe on a pole, a line marking a journey from Norway to Alaska, and beneath it is a dedication to Roald Amundsen, and his 'visit paid to Eagle'. Amundsen arrived here at noon on 5 December 1905, after a month and a half on a dogsled, in search of a telegraph station. Eagle had got a telegraph station in 1903. Amundsen had a message for the world that would not wait: he had found the route to the Pacific through the Arctic Ocean, the long sought-for Northwest Passage.

'Alaska, The Last Frontier' it says on the number plate of every truck in town. This here, I will be told time and time again, by gold prospectors, dog mushers, oil workers, fly fishers, those born here and those moved in, this is the last great wilderness on earth.

In 1890 the Superintendent of the Census declared that the

United States no longer had sufficient unsettled land that it could be said to have a frontier line, and that as such the notion of a frontier would cease to feature in the census. The grey had been erased from the map. And with that statement, wrote historian Frederik Turner in 1893, the first epoch of American history came to a halt. For Turner, the interplay between civilization and the wilderness had been integral in shaping the nascent country's character, in the 'Americanization' of its people. The frontier was where refined European habits hit up against the untamed land, the liminal place where wilderness untethered the civilized mind:

> It finds him a European in dress, industries, tools, modes of travel, and thought. It takes him from the railroad car and puts him in the birch canoe. It strips off the garments of civilization and arrays him in the hunting shirt and the moccasin. It puts him in the log cabin of the Cherokee and Iroquois and runs an Indian palisade around him. Before long he has gone to planting Indian corn and plowing with a sharp stick, he shouts the war cry and takes the scalp in orthodox Indian fashion. In short, at the frontier the environment is at first too strong for the man.

First, man wanders in the wilderness, but then he shall overcome. For over two hundred years the pioneers had pushed west across the continent, domesticating and converting as they went. To those who entered this 'simple, inert continent', the Indigenous people comprised part of the savage earth, as though formed of its very clay, an addendum to the wilderness to be subjugated with the rest. In 1845, John L. O'Sullivan, the political columnist, articulated the mission: 'the right of our manifest destiny to over spread and to possess the whole of the continent which Providence has given us'. The land was brought under the plough, the Indian brought under the gun. And the American pushed on. It was the pioneer spirit that brought him here; he cannot find peace within his homestead. There is an 'expansive power' inherent in him, wrote Turner, a 'restless nervous energy', and as soon as he feels hemmed in by other settlers he breaks for the woods once more. It is the original curse of man, a yearning for both freedom and familiarity. And so he continues, pushing forever west, until the final decades of the nineteenth century, when he hits up against the Pacific Ocean, and he stops. He stands there looking out over the water, three thousand miles of country at his back, and he flexes his muscles, and he thinks, what next? Where next?

I paddle out of Eagle, and then drift through ten miles of space. There is a beautiful monotony to the infinity of trees, the same conservative palette of greens and browns and greys and blues. There are flashes of colour, lupines and vetch, in the dark scrub of the banks. The weather changes, changes. The wind, now from off the water, now from out the hills. The mosquitoes flail about, legs dangling, probing at my clothes. The conveyor of the river, boiling, swirling, eddying, sweeping me along.

The current is beginning to slacken. I am only 850 feet above sea level, a sea which is still well over a thousand miles away. The land is softening, the slopes deciduous, the hills are less severe. Though still at times there are great bluffs that I drift past, slopes of talus with a mist of dust rising from them, slopes so steep they cannot support their own weight, and a fine shale rains down into the water.

Alaska is one-fifth the size of the rest of America (which Alaskans call, like members of a cult, the Outside). It is the size of France, Germany, Italy, and the British Isles combined. The population of France, Germany, Italy, and the British Isles combined is a little more than 277 million. Alaska has a population of 741,894 (as of 2016), which allows each citizen roughly a square mile each, although with 400,000 in the Anchorage metropolitan area the share for the rest goes up considerably. There is a joke Alaskans like to tell: the good thing about Anchorage, they say, is that it's only twenty minutes from Alaska.

Like the other animals, the people have their territories out here. A combination of personality and resources means that cabins are rarely in sight of one another. Each homestead requires an eddy for a fish net, some decent hunting grounds, and enough space for a man to think. I can see Andy Bassich's place for an hour before I get there.

'Twelve miles out of town, on the left' were his directions. 'Watch out for the eddy. Don't come till after working hours.'

I swing into the eddy, a lurch from rushing river into backflow that can flip a boat if you're not careful, and moor up. Dogs turn circles around my feet and escort me, yelping, to the house. Andy is sitting on the front porch with a beer. He is in his late fifties, a young face, grey hair, some cartoon figure on his T-shirt. I am amazed that this arrangement made a week ago from Canada on a broken phone line has resulted in an actual meeting. He gives me a brief nod, as though I have just walked in off the street.

Everyone does the bush in their own particular way. 'When I first got out here, there was a guy living down at Coal Creek,' Andy says, coming back onto the verandah and handing me a chilled glass of homemade beer. 'All he did all winter was sit in front his window. He'd read, and he'd read, and he'd smoke some dope, and he'd read. He'd do as little work as he could to prevent himself freezing to death. Maybe he'd go shoot a moose or a bear if it came into that window. He didn't have a pot to piss in. And he was happy as a clam. It wouldn't have made me happy. But it worked for him. Me, I like to work. I like to create stuff.'

Andy's setup is what happens when an architect has removed from the equation the slightest constraint of space. Where other people have rooms, Andy has buildings. There is a sauna. There are guest cabins. There is a workshop and a toolshed and a cache and several other buildings whose purpose remains a mystery. He lives in a two-storey construction with a porch, and he is at work building a new place for himself on the other side of the yard, piles of spruce planks stacked beside it that he has ripped from logs cut from the forest. There are drying racks, where hundreds of brown and brittle chum clack together in the river's wind, and greenhouses, where starts of cabbages and tomatoes and chillies and marijuana jostle for space in their short and frantic season. There are nest boxes for the swallows; they keep the mosquitoes down. There is one shelter dedicated to the building of boats, the shell of a kayak, steam-bent and wood-framed, suspended from the rafters. On a far-off bit of his land, he is setting up some yurts.

Back in the old days it had been busy out here, if one's concept of busy was fairly fluid, much distorted from years of remote bush living. But there had been families, and youngsters like himself. It might take you a couple of weeks for a round trip to Eagle, lining the canoes upriver, a few days in town getting some business done, getting some drinking done, and then the float back down. People would keep an eye on things while you were away, dogsit, babysit. When there was a new cabin to be built, everyone pitched in. Now, Andy says, the land feels empty. Young people didn't move here any more. It was just the old codgers like himself, getting older every year, and they were a dying breed. Old Dick Cook was taken by the river a few years ago, and he got lucky. Most of them would end up forgotten in a retirement home someplace, no one to look after them but some pretty nurse, any family they might have had long since moved out of state. Well fuck that, Andy reckoned. When his time came he planned to build a raft and float off down the river.

'The young people, they come out here and they see what

I've got and they expect to have it right away,' he says. 'They don't know that this takes time. They don't see the billions of calories I've used up building this place. Even clearing a little patch of land in the forest, without machines, that's a lot of work.'

He smiles and sips at his beer. 'They don't want to put in all that effort if they don't own the land. But no one owns the land. Just because I've got a bit of paper that says I do, it's not mine when I die. The payoff is in the experience gained. But they don't see it like that.'

Andy came from Maryland. At twenty-two, whilst training in construction management, his grandma gave him the best piece of advice he reckons he ever got.

'You've got time later to get a career, to get money,' she said. 'Right now is the time to have fun.'

He packed in college and pointed his new jeep north. There was a place on the map called Eagle, and how could you resist? But by the time he crossed into Alaska he was broke, and he ended up in Fairbanks. It wasn't until years later, on a canoe trip with a buddy, that he finally made it here. He arrived in Eagle on the 3rd of July. He stuck around for the 4th, and then he stayed for ever.

Andy's story is not a rare one. According to the 2010 census, only 39 per cent of those living in Alaska were born here. Whether genuine infatuation or some geographical manifestation of Stockholm Syndrome, it means that the majority of Alaskans have moved to Alaska, and the majority have stayed. Few people I meet are ambivalent about it. Forty degrees below does not permit ambivalence. People come on a posting with the military, people come for the rumoured riches. People come to be as far as possible from their family and yet still live in America. And some people, like Andy, come because it is a scratch for the frontier itch, a place where one could live beyond The Man.

Andy got hold of a cabin in what was optimistically called town. One winter, out mushing on the frozen river, he pulled in at the spot where we sit now.

'I just felt that I had lived here before,' he says. 'I had the most overwhelming sense that I was home.'

Across the water is Calico Bluff, strata of Mississippian and Pennsylvanian limestone enfolded with layers of black shale coated in sulphur, compressed by the force of the earth into an anticline so tight that it looks like a fingerprint, or a church door, whorls of yellow and grey and green. To the east are the mountains of Canada; ahead, a bar of spruce, scraps of mist caught on their tops, burning off in the late sun. A raven, clucking like a chicken on the bank, hopping about with rakish swagger in its baggy shorts. Only later, after settling in, did Andy discover that this is the second most important archaeological site in Alaska. His dogs are forever digging up arrowheads. People have been having an overwhelming sense of home here for many thousands of years.

There is king salmon on the grill (last year's, from out of the deep freeze), and caribou steaks. Fat drops of rain fall from a summer's sky and hiss when they hit the coals. He picks a couple of handfuls of lamb's quarters that are growing wild behind the dog yard.

'I've realized one thing over the years,' he says, flipping the steaks. 'Every year I bite off more than I can chew. But every year all the important stuff gets done. Every morning I make a cup of coffee and sit out on the front step and say, what do I want to do today? What are the priorities? There's not many people get to do that.'

The rain worsens and the mosquitoes are biting. We eat inside. The salmon is fat and juicy and delicious, its oils dribble down my chin. Andy tells me that this frozen fish is nothing compared to the fresh stuff, but to me, used to flaccid fillets from supermarket chillers, it is exquisite. Hand-carved duck decoys sit on hand-carved shelves, alongside a hand-carved rack of nine guns. Dried chillies and beaver pelts hang in the windows. Dr Seuss leans against Bill O'Reilly on the bookshelves, beside *How to Build a Sailboat*, beside *Forgive Instantly and Live Free*. A photograph of a favourite dog on the wall, with shockingly wild blue eyes.

'Security,' Andy says. 'Being able to feed yourself, being able to keep yourself warm. That's the only security there is.'

Once that security was salmon, a bounty in what Andy describes as hungry land. Like true fishermen everywhere, he speaks of salmon whose sizes bordered on the mythic. Stuck to the refrigerator is the same photograph that Richard Dewhurst had shown me, of Richard's dad holding up a monster fish.

'In the seventies it would take three people to drag a fish into a boat. Three. But those days are gone,' says Andy.

A king like that, a big female bound for Canada, might lay twenty thousand eggs, and would have the strength to dig redds that are deeper and more secure than the smaller fish can manage. Salmon live a lottery with outstandingly bad odds, a lifetime mortality rate of over 99.9 per cent, and they need all the help they can get.

'Burbot eat 'em, seagulls eat 'em, pollock eat 'em, everything eats 'em,' he says. 'They run the gauntlet. So if you can get two back, you're doing pretty well.'

In boom years, productivity has been up to five salmon returning for each spawning adult, but recently, as the size of fish decline, and smaller fish lay fewer eggs, it is more like one to one.

'That's what destroyed the fisheries,' he says. 'This is just basic math. You've been to college, you figure it out.'

In the past few years more conservative fishing, including outright bans, and a more rigorous approach to management, have hiked the run numbers back up. But numbers alone, as fisheries managers are only coming to realize quite late in the day, tell at best a superficial story. You could let a hundred thousand fish across the border with much fanfare, but if the eggs those hundred thousand fish produce are only going to result in no more than a hundred thousand fish returning, you have an extremely vulnerable system. There must be some excess built in to ensure resilience. Otherwise any overfishing, any environmental wobble, could decimate the run between one year and the next. Increasingly, biologists are speaking of ASL – age, sex and length – as the qualifiers that determine a healthy population. They talk of run quality, not quantity. And going by those measurements, the Yukon Chinook is looking very unhealthy indeed.

'Once the guys like me are dead and gone then there's no memory of what the good years used to be like,' says Andy. 'There's young guys now, they've never even seen 'em. To them twenty pounds is a big fish. Christ almighty, ten years ago they were the ones we were throwing back.'

It's not as if there was no warning. In 1976 a study by Canada's Fisheries and Marine Service had advised that 'greater use of gill nets and those of large mesh size, six inches or greater, may result in significantly greater exploitation of female fish [which tend to be bigger than the males]' and 'may threaten the (Chinook) salmon population with extinction'. Yet before 2010, mesh size for gill nets was unregulated; the Yukon may have been the last large-mesh commercial salmon fishery in the world.

Gill nets are set to span a section of the river, and will snare the gills of fish that are the same size as the net's mesh, whilst larger fish bounce off and smaller fish pass through. This is known as the Goldilocks effect: not too hot, not too cold, just right. By using mesh of up to ten inches, fishermen were deliberately targeting the biggest salmon, the strong females full to the gills with eggs, the cornerstone of the run. The largest fish, the seven-year-olds that spend five years in the ocean, are about

70 per cent female; fish that spend only one or two years at sea are virtually *all* males. It doesn't take long, selecting for the bigger fish, to drastically alter the sex ratio on the spawning grounds. Today fewer than one per cent of fish in the Yukon River are seven-years-olds. Only 14 per cent are six-year-olds.

Since 2004, Andy and a handful of others had lobbied, every year, for a reduction in mesh size to six inches. Finally, in 2010, the US Board of Fisheries brought it under regulation. But mesh size was set at seven and a half inches, rather than the six they had been pushing for. With the big fish now fished out, all seven and a half inch was going to do was select for the biggest that remained. Andy wanted six inches or nothing at all. And it was then that the runs really crashed.

'It hurt a lot of people,' he says. 'But we have a chance now to turn it around. It takes getting kicked in the ass.'

What is needed, he says, is to get the size of the fish back up. When a population is put under significant pressure, by overfishing, say, or reduced food in the oceans, they will reach sexual maturity earlier and breed at a smaller size, much like a flower bolting. Fishermen are now seeing a predominance of young males, known as jacks, as well as mature male parr, which have never been to sea. Often they adopt the markings of the females to sneak past the larger males in an attempt to fertilize the hen's eggs. Small fish, with their modest genes, beget small fish.

'The big fuckers come in last,' says Andy. 'I see some pretty damn big fish in August. I only care about how many females go across the border.'

Andy believes that to redress the altered genetics of the run will take eight to ten cycles of fish, that is to say, at least fifty years. Others estimate as long as a century. Humans are the greatest evolutionary force on earth: however long it took to make the changes they created, it will take longer for nature to redress them. It doesn't take much to get a spike in run numbers, just a couple of years of closures, but to turn around productivity is a much longer-term commitment. With a new fisheries manager in the post every three to four years, it is hard to maintain policy, and when fishermen see fish passing, it is hard to convince them not to fish. Politicians want to be popular, like most of us. Already, with numbers for this year forecast to be a little up on last year, people are itching to put their nets in for the kings.

'There's nothing about salmon management that's year to year,' Andy says. 'It's five years, ten years. And if you think like that it's a hard sell. Being a fish and game manager, you gotta wear kevlar, man.'

'And so how long would you leave the ban on kings?' I ask him.

He puts down his fork, fixes me in the eye.

'I'd leave it for ever,' he says.

Some salmon stray. For all their rootedness to place, for all the hard work and habit of their ancestors, a few per cent of each generation will stick two fingers to tradition and seek out new spawning grounds. For any species with ambition, this is vital. Most of the Pacific salmon's current range was glaciated fifteen thousand years ago: if straying were not wired into a subsection of the population, Alaska would have remained forever fishless. It was during this Last Glacial Maximum, with the sea level 125 metres lower than today, that several thousand humans first walked east across the land bridge onto the virgin continent. Yup'ik people, with common genetic descendants, walked east from Asia across the land bridge, with the shared technologies of kayaks and harpoons and waterproof parkas made from seal guts. The Yukon, the last river in the Americas to be explored by Europeans, the first to be travelled by Pleistocene hunters.

Salmon stray because many things can go wrong with a creek. A landslide, a logjam, a flood, a dam, a spill, a gold-dredging operation: all can make once viable spawning grounds inaccessible, unusable. The eruption of Mount St Helens in Washington in 1980 boiled and buried alive all fish in the north fork of the Toutle River. At least five times in the past three thousand years the forward surge of Lowell Glacier has dammed the Alsek River in Canada, most recently in 1909. Only the returning offspring that strayed that year would have survived, and the repopulation of the Alsek's upper watershed, since the glacier's retreat, has been entirely the work of strays from other runs. When dams are removed the fish come back, and quickly. Rivers once so polluted they were flammable have been recolonized since their waters have been cleaned. On the maps the salmon hold, these rivers are the true frontiers, spacious and unpeopled. Thomas Quinn, salmon biologist, sees straying as a genetic disposition, but remains uncertain exactly what it is the fish are disposed to.

'Are some fish "programmed" to stray,' he wonders, 'or do they fail to home? That is, do strays identify the home stream but then go elsewhere, or do they have poorer memory or sensory capabilities and so stray out of ignorance?' That is to say, are they genetically bold, or are they genetically lost?

The Chinook's historic range once grazed the Tropic of Cancer and went far north of the Arctic Circle, well in excess of three thousand miles. They spanned from the Ventura River, in southern California, to Kotzebue Sound on Alaska's west coast, and on the western Pacific Rim from Japan's Hokkaido island as far north as the East Siberian Sea. Today there are also viable wild populations south of the equator, in Chile, Argentina and New Zealand, a mix of escapees from salmon farms and introductions by homesick anglers. Over the past few decades these southern fish have inserted themselves into the ecosystems of a hemisphere that has no knowledge of them. The Patagonian landscape is not dissimilar to the salmon's traditional habitat, and they are colonizing rapidly. No other anadromous fish has such a range in South America; the salmon already span fourteen degrees of latitude, and it seems unlikely much will slow them down until they push up against tropical temperatures they cannot handle.

In the northern hemisphere, too, their range is expanding. As oceans warm, the salmon, which rely on oxygen-rich cold water (carnivores need plenty of oxygen), are wandering further north into the Arctic, and they are showing not only behavioural but also genetic adaptations. All five species of North America's Pacific salmon have been found along Alaska and Canada's northern coasts, harvested as far east as the western villages of Nunavut. There are communities catching these fish who have no idea how to cook them or how to preserve them, who have no word for them in their language. It is not yet known if there are established breeding populations in the Arctic; the MacKenzie, the major river system flowing north into the Beaufort Sea, is even more massive than the Yukon, and if there is an advance guard of spawners they are a needle in a haystack. But Karen Dunmall, of the Canadian Department of Fisheries and Oceans, has devoted much of her recent research to demonstrating that conditions in parts of the MacKenzie are favourable enough that at least some species *could* breed in this new territory. Concurrently, Atlantic salmon have been advancing around the tip of Quebec and into Hudson Bay. In 2012 one was caught in the Clyde River, on Baffin Island, in eastern Nunavut, putting the frontiers of the Atlantic and the Pacific salmon at just over a thousand miles apart.

There is no reason not to suppose that in the not too distant future an Inuit man will head out to check his net and find an Atlantic and a Pacific salmon entangled side by side; that these two genera of *Salmonidae*, unknown to each other for twenty million years, will navigate the Northwest Passage a little more than a century after Amundsen made it through, and come face to face once more.

'And *won't* that be an exciting day!' says Dunmall.

I ask her how soon she thinks that it will happen, but she will not be drawn into speculation: she is a scientist after all. But when it does happen it will go down in the annals of salmon history as a first contact as significant as that of Columbus or Cook. Whether these creatures are able to cohabit, or whether it will be at the expense of the eradication of one, past histories can only give an indication.

After dinner, I follow Andy out to feed his dogs. Each has its own kennel, a plywood shack, and each is tethered to a pole out front by a length of chain that allows for an orbit of about fifteen feet diameter. Each kennel is so arranged that the orbit of one never intersects the orbit of another, the ground trampled to bare earth along the entire arc of their trajectories. Dogs lie in the shade, dogs loll on the roofs of the kennels, dogs walk in circles, first one way, then the other. One paces with its bowl clamped in its jaw, like a lip plate. They are frantic with anticipation, chains rattling like convicts. They whine when they see him, manic with love. Andy makes each one sit before he feeds it, and they do, forcing their heads against him. He hacks at the dried chum on a block with an axe: only one in ten of the chum that come this far are still firm enough for human consumption, but dogs are not so choosy. He drops some pieces in each bowl, and the dogs crack on them like bones. Some days he releases them all and drives his snowmachine upriver, and the dogs run the banks beside him. Most people don't keep so many dogs now; it was largely a cultural eccentricity of a few old white folks. But you can't love a snowmachine. And you can't eat a snowmachine when you break down out in the bush.

I look about the yard. Andy did not build this place alone,

although to hear him speak you would not know it. And he did not rebuild it alone either, for everything here has been rebuilt since the flood of 2009. But you would know that if you were one of the several million viewers who tune in to *Life Below Zero* every week, the reality TV show that Andy and his ex-partner Kate Rorke star in, along with a handful of others who live along the line of the Arctic Circle, 'subsisting off the rugged Alaskan Bush'. You might well have debated the intricacies of their lives across the breadth of social media. Having a camera crew dogging their every move, and their love life play out on national television, each partner's every slight analysed by a million relationship experts, was a strange payoff for rural remoteness.

There was a time when fishing had paid Andy's bills, until the commercial ban was implemented. For several years he captained the *Yukon Queen*, a catamaran that ran a daily service between Dawson and Eagle, before it was axed in 2012 because of the destruction its turbines caused to salmon fry. You needed money to live in the bush these days, however off-grid you were. That was the mundane reality of reality TV. You needed gas for the tank, tin for the roofs, coffee, a hundred things that no one used to need. Just like the furs, just like the gold, the guys on the shows had a resource that those outside Alaska wanted. They were selling the life, the dream, the wild, and the Discovery Channel and National Geographic had tapped a gigantic market for it: *Ice Road Truckers*, *Deadliest Catch*, *Bering Sea Gold*, *Alaska's Wild Gourmet*, *Alaskan Bush People*, *Flying Wild Alaska*, *Sarah Palin's Alaska*, *The Last Alaskans*, *Alaska: The Last Frontier*. Along the river I will meet no fewer than four men who make a large part of their income from being on reality TV. It seems quite possible that Alaska has the highest ratio of television celebrities in the world.

Kate Rorke left Eagle, and Andy, and *Life Below Zero*, in the Christmas of 2014. Fans of the show were shocked, if not surprised. She lives now on an island off Vancouver Island. There is a blink-and-you-miss-it downtown, but it is not the Alaskan bush. When we speak on the phone, she is much curtailed in what she can say. There are lawyers involved, both Andy's and those at National Geographic, which owns the show.

Kate had met Andy in Dawson whilst on holiday. In 2004 she went back to live with him. Andy was in Eagle then, with just a wall tent and an outhouse and a sauna on the land at Calico Bluff, a partially built cabin. Kate had never longed for Alaska with a gravitational certainty, had never had a sense that she had been born in the wrong place, as many other incomers speak of it. But she had lived in small towns, and she knew that she was comfortable in her own skin.

'It's scarier in the city than it is in the bush, I can guarantee you that,' she says. 'You know how an animal's going to react. You don't know how people are.'

In an interview with National Geographic she said that she missed nothing but her family and high heels. Andy and Kate got the cabin up. She loved the simplicity, and she loved what she was learning. But as time went on, and more buildings went up, more solar panels, more satellite dishes, internet connections, she began to long for how things had been. They would visit neighbours in their one-room cabins, and Kate would come away pining for the uncomplicated nature of their lives. People back home would ask her what she did all day.

'Well I don't sit down,' she would reply.

And it wasn't just the time, it was the expense. Kate poured in her money, because Andy didn't have any. But back then, she thought she was building her home. A home she would lose twice.

The break-up of the Yukon is always a dangerous time. Throughout winter the river is a solid highway, but in the spring it becomes precarious and fickle, and for several weeks travel is impossible. Then one day, without much warning, the ice will shift and part and make off for the sea. Along the river's length people come outside to watch a million tonnes of ice moving along at ten miles an hour, as though the very land were walking. More tangibly than anything, it spells the end of winter's grip. The ice shifts and piles floe upon floe, screaming and grinding as it goes. In the spring, a tripod is erected on the ice outside of Dawson, connected by a wire to a clock that is tripped and stops when the ice beneath goes out. Since 1896, when gold miners with money to waste and time to kill began gambling on the moment as a good way to see in spring, the time of each event has been recorded in a ledger kept in town. The first time break-up occurred before 1 May was 1940, and then again in 1941. There was not another April break-up until 1989; since then it has occurred eight times. In 2016 the Yukon smashed the record by more than five days, going out on 23 April at 11.15 a.m.

In 2009 the ice went out from Dawson at twenty-two minutes past midday on 3 May. Kyla McArthur won $3,500. At seven o'clock on the evening of 4 May, Andy and Kate were a hundred miles downriver on the bank at Calico Bluff, watching the landscape awake from hibernation. Kate had never missed a break-up; it was a beautiful time of year. Everything was moving quickly. Four of the past five days had been hotter than ever for the time of year. An almost instinctive sense, formed from many years of observation, suggested that this break-up could be out of the ordinary. They had taken the precaution of hauling the guest cabin, with its river views, seventy feet further up the bank.

The ice was going past like magma, the water ebbing and flowing like something breathing, and they saw that on its surges it was already breaking the Yukon's banks. Downriver at Six-Mile Bend a great wall of ice was forming – 'like a Trump wall, I call it now,' says Kate – and the river, released from its winter cage, was building up behind it. The water was rising steadily, coming up way too high. They watched their picnic table float off downriver, and dog bowls, garden tools, oil drums, logs and tarps and bits of lumber. Then the water came up six feet in five minutes.

Andy hurried about the property in his waders, the ice-cold river already at his waist. It was rushing, seven or eight knots of current, plucking at his feet. Kate was in their cabin, trying to haul what she could upstairs. Andy lashed his two canoes onto one of the skiffs as outriggers, and pushed the raft in front to keep him steady. He moved across the yard to where their twenty-four dogs were perched on the roofs of their kennels, observing their world turned upside down. He unclipped them from the chains that held them. Some leapt for the raft. Some leapt for the water and paddled for the porch, breathing hard through their noses, heads straining against the tide. Vixen was there, forgotten, dog-paddling on her leash. Some made it to the decking; others were caught by the current and carried off into the woods. They hauled themselves onto timber piles and styrofoam and snowmachines and anything that still stood above the waterline. Delete was trapped in the puppy pen, fighting against the tide. Andy could see one of them, Jack, maybe, perched on the barbecue that had lodged between two trees, staring back at him, barking, and then the barbecue toppled over.

Ten o'clock, and the sun going down. As darkness came on the surge abated, and as the water calmed, dogs paddled out of the woods, heaved themselves onto the porch and shook the river off. Andy counted them. Twenty-four, thank God. But he knew enough to know it wasn't over. Moving about in a canoe, he and Kate made their preparations. From the cabin they gathered food and clothes and blankets and whatever else they'd need to make it through the night. He got both of the skiffs up to the porch, and lashed a canoe alongside each of them for more stability. The water was inching up again and had already reached the decking when all of a sudden it burst up from out of the house's cellar. It rose inside the cabin, to their ankles, their shins, their thighs, as they rifled through their possessions, taking instant stock of what they needed and what they could lose for ever. Upstairs, Kate got Eagle on the marine band radio; hell was breaking loose there too.

'If you don't hear from us by morning,' she said, 'then we need saving.'

They forced open the door and waded out into the darkness. Andy hefted the dogs into the bigger of the two skiffs and clipped each to a thwart. Kate climbed into the smaller boat, tethered to a birch tree, and Andy climbed into the other, tied onto one of the porch's uprights.

Two humans and twenty-four sled dogs, floating in the dark: all that was needed to repopulate Alaska when the floodwaters receded. It was well below freezing. The old-timers used to gauge the cold by how many dogs were needed in the tent at night to keep warm. A four-dog night. An eight-dog night. They kept an eye on the level of the water with their headlamps. It had reached the doorknob of the cabin. Andy tried to calculate what that meant: maybe a rise in the water level of thirty feet? Bergs drifted past in the torch's beam and passed back out into the night. All the accumulation of their life, suspended in the darkness. They kept vigil and spoke little. There wasn't much to say. By the time that the sun cracked the sky, they had done a lot of thinking.

In the dawn they could see that the waters were beginning to go out. But they were going out terribly quickly, rushing to fill some vacuum downriver where the ice jam must have burst. The whitewater surging, tossing the boats about, worrying at the lashings. Andy saw his Rotavator heading for the river: two hundred and fifty pounds of machinery, tumbling over and over like a piece of drift. The sauna nearly hit them before the currents shifted, and it rushed past beyond Kate's bow. Trees cracked and snapped, metal howled. A shed slammed against the cabin, the building shaking to its foundations. As the water level fell, Kate's rope held where it was tethered to the birch tree, and by the time she noticed they were already listing and the canoe was swamping with water. Kate stood in the bow, hammering with her fist against the rope to try and yank it down, but it was jammed, and the whole contraption was starting to invert. She needed to cut the canoe loose before it pulled the whole thing under, but there wasn't enough leeway in the painter to get to the stern to cut it. She was trapped. Either the canoe would take her down or the stuck painter would flip her. She looked over at Andy.

'And I just said goodbye,' she says.

But somehow Andy got from where he was across to Kate, and cut the ropes that held the canoe to her boat. The canoe shot forth as from the barrel of a gun.

'The only way I can describe it when it hit that tree,' says Kate, 'it looked like tissue paper on a rose bush. Just crushed and collapsed, with such force that it buckled it in two.'

Then Andy's skiff tipped, flipped on its side, and went under. The water took hold of it and pulled, dragging the boat beneath the chop and all the dogs down with it. Some went under, some held by their teeth to the gunwales, eyes enormous, and Andy stood there, one foot up on the porch and the other on the submerged bow, reaching for one dog at a time and cutting its tethers and slinging it by the scruff onto the decking. They sat where they fell, their nails dug into the wood, shaking and wild-eyed and mournful. Then Andy's canoe came loose. The five dogs in it leapt for the decking. Three made it, two didn't, and he watched as Iceberg and Ouzo swirled off into the river. The water kept on rushing out; finally, the boats grounded. It was only then that he found Skipper, still tied down in the stern, shrunken, bedraggled, and dead.

The yard was a slop of mud and debris. The dogs ran circles in the mire. They called them, rounding them up. Remington, Donner, Snickers, Amber, Control, Jack. When he called for Iceberg and Ouzo, they bounded from the forest. And Iceberg had only got over her fear of water that past spring. Skipper was the only one that Andy had lost. One moment of oversight, one stupid fucking mistake. Chunks of ice as big as trucks sat about the yard. They were starting to assess the damage when Kate looked upriver and saw the unthinkable: another surge was coming.

She didn't think that they could last out another one. The cabin was much weakened. Many of the trees on the upriver side that had broken the water's initial force were down. Andy wanted to head up the bluff; he didn't know then that they were now on an island. Kate insisted they needed rescue.

'Sometimes,' she tells me, 'toughing it out is not the best idea.'

She tried Eagle on the radio but got no answer. But she managed to get their neighbours, Wayne and Scarlet, on the satphone, and they passed on a message to the Park Service. The Park Service would try and scramble a chopper, but they needed all obstructions removed from within a fifty-foot radius of where the pilot would touch down, which was going to have to be the cabin roof. Andy grabbed a saw.

In the next hour he felled thirty trees. He worked like a man possessed, striding through the flood. He was done and up on the roof by the time he heard the helicopter, the dogs back in the boat, the water at the decking. A Raven, a two-seater, growing larger from the south-west. He was so grateful to these people, so grateful to be Alaskan.

Rick Swisher, the pilot, held the helicopter steady, one skid against the cabin's shingles, while Andy moved about under the blast of the rotors, the river whipped up, helping Kate onto the roof.

'I need to know if you're going to take the dogs,' Kate said to Rick once she was inside. 'Because if you're not going to take all the dogs, I'm not leaving.'

'We're taking all the dogs,' said Rick. 'My wife said I wasn't to come back unless we do.'

Kate pointed to a flat piece of ice, a few yards out from the porch, that might work as a landing. Rick hovered over it, tried it tentatively, once, twice, and then committed. The ice held. One step at a time, he told himself.

Andy paddled out in the canoe with the first three dogs and slung them one by one into the back of the Raven, one hand on the collar, one hand under the belly. Kate tied them down, the seatbelts through their harnesses.

'Make sure they're in tight,' said Rick. 'Because if they come up here we're dead.'

The helicopter lifted off as Andy paddled back for the next three, and Rick made for the Park Service airstrip sixty miles to the north-west. As they rose, Kate looked back upriver. Any landmarks she knew were underwater, the river vast and blurred into the land so that they seemed to fly over some unknown sea, thick with ice, with no shape or definition. Calico was the only dime of land still visible. Eagle was devastated. Blocks of ice as big as cabins where the cabins she knew once were. The customs house pitched at an angle, the general store gone entirely. The streets were all water, reflecting back the sky, and debris, everyone's stuff, bobbing on the swell. The boat that Andy captained, *The Yukon Queen*, was wedged in a stand of forest, a few hundred feet inland.

That was eight years ago. Kate still has nightmares about the flood. She shakes when she speaks about it. But she never lost faith in the river.

'National Geographic would ask me what I was afraid of,' she tells me. 'And I would tell them nothing. I don't carry a gun, I don't carry a knife. I have a feeling that the animals know I'm not a threat. If you're gonna live in the bush and you're gonna live in fear, you don't belong there.'

She misses Alaska every day. She walked out; she tells me that leaving was a matter of survival. She left the dogs behind, and she had been midwife to almost all of them.

'I can't have a dog to this day,' she says. 'It was like I had my entire family wiped out in a heartbeat.'

She watches the dog-mushing races on cable every year, she calls her friends in Eagle once a week. She has thought about moving back to Dawson, into a cabin, but at sixty she knows that she's only got a few years left where she could manage that way of living by herself.

'I loved it there,' she says. 'I loved the lifestyle. And I loved the land. But I couldn't love the man he really is.'

There is an old Alaskan adage amongst the women, frequently rolled out, predicated on the fact that the state has the highest ratio of males to females in the United States. Kate quotes it to me now: the odds are good, but the goods are odd.

'A lot of these people in the bush,' she says, 'they're there for a reason. And some should be by themselves. Some people don't choose the solitary lifestyle because they want to live a certain way. They choose it for a different reason. Society's hard for them.'

And yet, Kate does not regret it.

'It's a sad story,' she says, 'but it's a good story in a way. I can't change the way it was, but no one can take that experience from me. It's a place I think everybody should experience at some point in time in their life. It's scary, in a magical kind of a way, and it's something that most people will never see. In a small apartment building in downtown New York, they have no idea what space and fresh air and freedom feels like.'

Andy and I look out at the river, in the buttery dusk. The dogs are fed. The sun is going down, and the Yukon looks benign tonight in the pastel-coloured half-light. There are clues, if you know where to look for them. A patch bare of trees, where the swallows dart. A high-water mark. They started rebuilding, straightaway, right where they had been before. This is home, as simple as that. Andy was assured that such a flood was a once-in-a-lifetime event. But the definition of once-in-a-lifetime is not what it used to be.

'There's been three floods close to that one since 2009,' he says. 'And *apparently* climate change isn't happening. That's what the Republicans say.'

None of those three touched Calico, not quite. But there is a sense of when, and not if. The frontier, pushing back. I ask him what he will do if there is another flood. Andy looks around the yard, shakes his head. From somewhere, a raven calls. It is a while until he speaks. He says that he would not have the strength to rebuild it once again.

I reach the Flats. The river unravels, frays like rope; vast wetlands spread out across the landscape. It is a week since I left Andy's, and for two days now it has been as though the Yukon has lost its purpose. It rambles across the plains, now two miles wide, now three, all urgency forgotten. It has the spacious feel of an estuary, but I am still more than a thousand miles from the sea. The map looks like a doodle, the islands and the lakes. Fairbanks, Alaska's second largest city, is a hundred miles to the south, and jet fighters out on training runs from Eielson Air Force Base turn doughnuts, flip on their backs and flip again, carving vapour trails into the sky. It is easier to conceive that they are giant metal birds than to imagine that there are cities and wars and multitudes somewhere away from here. In the distant hills back towards Canada, another wildfire is smouldering.

Willow and alder, alder and willow, willow and alder. The deciduous mark the shifting sandbars, the slower-growing black spruce mark the banks; it is the only way to distinguish them apart. It is so easy to get lost here. Even the best of the riverboat pilots must learn the new channels every season: the land is fluid also, different from the water only in its speed of flow. I follow what I believe to be the south bank to round a corner and see it was an island, miles long, and now there is another expanse of water, stretching over to the left. I aim for channels that seem like shortcuts and discover that I am paddling upriver. It is so easy to get lost here, but you would not stay lost for ever. You may take a slough that winds sluggish through the scrub for many miles; or run aground on some unexpected sandbar a few inches beneath the surface, a mile from either bank, so that as you get out to haul yourself off, it appears as though you are walking on the surface of the water; you may try and gauge direction by a sun that never sets; but if you stop paddling and sit back the river will take your boat and lead you there. It might detour, it might take you weeks longer than you had planned, but it knows where it is going.

The spruce is sculpted by the elements, bottlebrush scrawny, topiaryed by the weather. This boreal forest stretches over eight thousand miles in an unbroken line around the circumference of the globe: 30 per cent of the world's tree cover, four million square miles, the planet's single largest biome. A broad, evergreen brush-stroke that encircles the north, running through North America, Scandinavia, Siberia, marking the band of the subarctic. Forests of moose, of lynx, of bear. Forests of thimbleberry, strawberry, nagoonberry, lowbush cranberry, highbush cranberry, watermelon berry, bunchberry, crowberry, huckleberry, blueberry, cloud-berry, bearberry, salmonberry. Forests home to many of the world's remaining hunter-gatherer societies, summers of wild-fires and perpetual light, and winters of fifty below.

Now the trees are rushing north, at the rate of a hundred metres a year. In 2008 there was a forest fire north of the Arctic Circle, where the local dialect had no word for wildfire. The Arctic regions have warmed at twice the global average, two degrees centigrade in the past sixty years, and the climate zones are shifting north ten times faster than the trees can migrate. Forest fires, racing through this dead wood, are burning hotter, longer, ranging further: 2015 was the second worst year for fires in Alaska's history, burning up an area the size of New Jersey (2004, even worse, was the size of Vermont). I paddle through old burns at times, battalions of dead spruce still standing, looking like diagrams of spinal cord. They do not rot in this dry air, but will stand like this until they fall. The burns make excellent morel picking. The boreal forest depends on fires for sporadic renewal, and so does the slow procession of recolonizers, the mycelia, then flowers like fireweed, then the first shoots of the aspen. It depends on them, but not like this, with their savageness and rate of spread and the scope of their destruction. Bugs like the spruce bark beetle and the aspen leaf miner, once killed off by brutal winters, are surviving and decimating forest; I can see their work from miles away, the pale turquoise that they turn the aspens' leaves. The forests are fragmenting: in Ontario and Quebec a gap of almost two hundred square miles has opened in this once unbroken circle. One day grasslands could take the boreal's place, or forests of oak and maple and black walnut, and the species that depend upon this vast habitat could be forced from the edge of the map.

Exhausted, wind-blown, I camp on a slop of beach. Thixotropic mud, oozing its liquid form. There is garbage everywhere, rusting machines and battered mattresses. Above the bank is a padlocked cabin, too buggy to contemplate. The tin roof pops in the heat with deep gong notes. 'Keep Out When Guner Not Home' says the sign, pockmarked with bullet holes. A confederate flag flaps in the wind and the raspberries are starting to ripen. There are bear prints on the shore, and there are great scratch marks through the goose squirt, but there is nowhere else to pitch a tent.

Of course, the possibility of bears has been the background chatter of my mind since I arrived here. Alaskan men trade in bear stories like men in other countries might speak of sexual conquests. They like to scare the tourists. It is not hard to do. Knowing that if I wrote out the food chain of the forest I would not place myself at the apex gives me a deep sense of connection to this land. I am accustomed to eating. Now, I can be eaten. The land feels more fleshed out, and as such this ecosystem seems more intact than any other I have known. Beside a creek I watch a grayling sucking flies from off the surface of the water, I watch a Wilson's warbler peck fruit from a high bush cranberry, I watch a bald eagle poised on the crown of a dead spruce, and I watch my back, starting at movements in the bush. And yet, the fact of bears is too much of a constant presence to be met consistently with fear. I feel, I assume, how herbivores must feel: jumpy and primed, primed for flight, but generally preoccupied by the daily grind of life.

The big mammals are rare. If I see one in a day it is a good day. But their very rareness seems to confer some further dignity both upon the country and upon their lives within it. A bear, or a moose, is emblematic of a much vaster, unseen wilderness. Like the salmon, it implies great space around it. A grizzly bear needs a territory, to itself, of perhaps one hundred square miles. California's state animal is a grizzly, but there are none there now but for the zoos (see also the bison, state animal of Kansas and Oklahoma). In the Lower 48, the range of the grizzly has been reduced by 97 per cent. In Britain we lost our last bear around the time of Jesus. Such is megafauna's seemingly inevitable trajectory as human population density increases. To know that bears are here makes me feel part of another time, not just another place. This far north, there is still enough space for them, and indeed, as the climate warms, they are expanding their range, pushing up against the polar bears. In 2006 a hunter shot a bear in northern Canada that was confirmed to be the first wild hybridization between a polar and a grizzly. It is known either as a grolar or a pizzly, depending on how terrifying you find its prospect. Since then, there have been two more confirmed cases. Unlike most hybridizations, their common ancestor is close enough that these grolars are able to breed.

I have seen bears, although as yet only black bears, not the grizzlies. A couple of days ago I watched one walking the bank, rooting for tubers of milk vetch. Delicate, almost feline, it kept pace with the canoe for perhaps a mile, a black so absolute it looked as though it had been punched out of the landscape. And one that I looked up to see one evening, as I sat by the fire, reading. It stood at the far end of the sandbar where I was camped, apparently unaware of me, perhaps a hundred yards away. The wind blew towards me; their eyesight is not good. I rummaged for the airhorn. It glanced at the noise, paused, and then moved nonchalantly off, travelling crosswise to the river across a series of islands until it made the quickest channel, where it flipped onto its back and floated off downriver, feet first, as though the *Jungle Book* were real. But these grizzly footprints – larger, heavier, altogether more solid – are right beside the tent, on what I now see is its trail from the bush down to the water.

I go for a walk. Further down the beach I kick up some of last year's king salmon bones in the brush, mummified beneath last winter's snows. Her jaw parted, rows of teeth bristling its length. The body bent into an S. Her backbone is frayed like worn wicker furniture, and with her paper skin pulled close about her she looks like some fairy-tale crone in cap and meagre shawl, shrunken and chilled to the bone. These bones will have been dragged here from the river by something, a bear, or a wolf, or an eagle. An otter even. There are over fifty different mammals that take nourishment from the kings. In southern British Columbia, the diet of the killer whale consists almost entirely of Chinook.

There is perhaps no image more iconic of the Alaskan wild than a bear haunch-deep in the turbulence, fielding leaping salmon like a goalkeeper. Bear density can be eighty times higher than what is typical where the salmon are plentiful. On a tributary of the Porcupine River drainage, bears have worn holloways between the caves they den in and the water, grooves in the earth that speak of the habits of millennia. Grizzlies must add 50 per cent to their body fat before they den for the winter.

They can get through forty salmon in eight hours; forty kilos of meat a day. Often they will take only the best, most calorific bits. They will bite off the skull to eat the brain, or hammer with a paw on a female's back, like someone with their change stuck in a vending machine, to see if they can squirt the eggs out. The amount the grizzlies put away before hibernation directly influences the number of cubs they will have next spring.

The remainder of the carcasses are abandoned, often left back in the forests where the bears go to eat them, away from the competition. Bald eagles, too, will hack apart a salmon for the parts they want and leave the rest for the ravens and the crows. An eagle well fed on salmon will have more chicks, and will hatch them earlier, giving them more time to fledge before the cold weather comes again. As the salmon descend the pecking order, they are feasted on by rodents, by smaller birds and beetles. The gulls on their migratory journeys south move from stream to stream to coincide with salmon runs, devouring rotting flesh and eggs. The insects that pick at the meat left on the bones bring in the songbirds and swallows and amphibians. Kneeling angelica times its flowering for ten days after the salmon arrive. This is when the blowflies emerge, to lay their eggs in the salmon carcasses. The blowflies need the angelica's nectar for the energy for reproduction, the angelica needs the blowflies as its pollinator. Their maggots break the carcasses down into the soil.

The carbon and nitrogen and phosphorus and fat that the salmon carry in their bodies from the oceans, up these old trade routes, are spread throughout the ecosystem, drawn up through the roots of trees towards their branches, restoring the fat of the land. Along some streams, the concentration of nitrogen and phosphorus in the soil can be so high that it exceeds that of commercially produced fertilizer. Spruce and willow grow up to three times faster along these streams. Up the Porcupine River, where the kings no longer come, the forests are dying. Imagine, if you like, the salmon swimming up the capillaries of the spruce and birch; it is not so far from the truth. Up to 70 per cent of the nitrogen in these forests had its origin in the sea. If you know the land well enough you can gauge the state of the salmon run by the fecundity of the forest.

The next morning, and no bears in the night, I wake to the cry of a peregrine. I watch it as the coffee boils. It is up the bluff, perched at the apex of a lightning-perished alder. Later, there are a pair of them above me, playing, the one knifing down, the other rolling to its back, talons upraised, and both of them veering before impact. Some ground bird chucking in fear of the spectacle. The sky is blue and they are sickles against it. There is a nest somewhere, I can hear the chicks. I wonder how far they can see up there, whether they see as far as the cities.

I paddle on. The wind carries on it the seeds of cottonwood, and the threads of spiders, the spiders attached and floating. It is near solstice when I cross into the Arctic Circle, and the sun spins circles overhead. High scrapes of cirrus, cat's claws, drag across a bottomless blue sky. Flocks of terns hang on the air, spun silver. Everyday I pass more abandoned fish camps, old tarps so sun-faded they are patchwork now, the lattice of their warp and weft exposed. In the seventies the tarps were all yellow. More recently they were green. Now, by and large, they are blue. Archaeologists will find this useful. The frame of an old bench, devoid of slats, perched on a rise for the best view of the river.

I see the bush planes of Fort Yukon before I find it, tucked away behind a scatter of islands on the north bank of the river. Scrawny dogs on chains pace the beach, their kennels cut from plastic water butts. I camp on one of the little islands opposite the town, intending to rest up for a few days. Every morning I paddle over and walk about. Fort Yukon is the biggest town on this stretch of river, population 550, almost entirely Athabascan, the native people who make up the majority of interior Alaska. It is laid out on a plan with total disregard for space and designed for total dependence upon vehicles: several miles across, the houses spread all over as though dropped from out of the sky. Without a truck, each visit I make entails long walks down dusty roads. Sometimes I flag down a quad-bike, what they call a four-wheeler here, and ride pillion, side-saddle, attempting to look jaunty. No two houses are the same. Many were built by their occupants, or by their fathers, or by their grandfathers, each one a manifestation of family history. The older ones are built from logs; the newer ones, funded by the tribe, from cheap ply panels, roofed with corrugated iron. They look entirely unsuited to winters of minus fifty. But there are only two seasons in Alaska, so the saying goes, winter and construction, and the quicker homes go up the better. They were designed for when gas for heating cost $1.25 a gallon, but now it can hit twelve dollars, especially towards the end of winter when supplies are running low. Most houses are raised up on stilts above the ground so that the building's heat will not thaw the permafrost beneath, and all have Arctic porches, a rough airlock at the entrance as a buffer from the winter's bitter cold. The yards are mounds of gathering debris: buoys and engines, fish nets and chainsaws, caribou skulls and moose racks, the ravens picking at them. Flowers push up around the snowmachines, parked up and awaiting their season. A chronology of the car industry from 1950 to the present; with nowhere to dispose of vehicles they sit where they died, to be cannibalized for parts. Improbably, one house has hanging baskets, and feeders for the birds.

Once there were four stores in town. Today there is just one.

The AC Value Center, it says over the door, *Since 1867*. I walk the aisles, looking at things, the novelty of shopping. It is cool, and music plays. There are candy bars for more than a buck, peanut butter for more than ten. There are rubber boots and outboard motors and power tools and grease. Mary Jackson stands behind the till. There are bags which are almost bruises beneath her eyes. She is from Pilot Station, she tells me, nine hundred miles away down the far end of the river, in Yup'ik Eskimo country. She ran an AC Store down that way, and then she was sent up here to manage this one. She moved, with her husband and her kids, two years ago. When she speaks of homesickness it makes you believe that it really could make people sick. It is worst, she says, during fishing season. A trip home is $1,700, three times more than my ticket cost from London to Whitehorse. She has not been home since she moved here. She shows me Facebook on her phone. There are posts from her family down in Pilot, totes full of salmon, some of which are kings, and fillets hung in strips on drying racks, the light behind them, and the colours are so vivid that I assume they have some kind of filter on them. She says that they do not.

'They're this rosy, peachy sunset orange,' she says. 'They're *so* good. When you smoke them they just *drip* with oil.'

Kings that enter the mouth can be as much as 24 per cent oil. Smokehouses frequently combust with the quantities of oil about. She flicks through more photos, of her siblings cutting fish, of her brother out in his boat, of a fish tart made with angelica. She taps to like each one. There are fish emojis, smiley faces.

'You'll be fishing soon,' I say.

'Not soon enough,' she says.

Even though Fort Yukon should be getting some openers this summer – the relaxations on restrictions during which time kings can be caught – Mary cannot find much enthusiasm for it.

'The kings at this end of the river, they aren't good,' she says. There is a wild sadness in her eyes. Her family is airfreighting some fillets to her: they should arrive any day now, before the swimmers. Everyone on the river will tell me the same thing: their fish, in their patch of river, the fish that they grew up with, is king salmon at its prime. The downriver people find the fish up here too dry. The people up here find the downriver king too oily, they say it makes them sick.

Mary invites me round for dinner. On the verandah her husband, Matt, from Idaho, is flipping burgers. They run the store together. He wears a trilby. On his hoodie it says: *All I care about is snowboarding, and like maybe three people, and beer.* I help Mary slice onions and tomatoes in the kitchen. Her kids sit around the room; it is the summer holidays, and there is not much to do. Her eldest, Rochelle, is back from Edgecumbe boarding school in Sitka. Edgekimoe, the white kids down there call it. Julius hammers at the Playstation. Cathy is outside, bouncing on the trampoline. Delaney is making cookies from scraps of paper and loose change, and Atreyu sits gurgling and snotty on the sofa, Woody from *Toy Story* in his hand, a T-shirt that says *I love my Papa*. Mary had Rochelle at seventeen. Rochelle is sixteen. She wants to go to Seattle to study media arts and animation. There are only one or two graduates who go on to college from Fort Yukon every year.

Mary is the tenth of fourteen siblings. They grew up in a cabin twenty-five feet by twelve at Pitkas Point, the newest baby in bed with her parents, the rest of them side by side on mattresses on the floor.

'There was a bathroom but we had no running water,' she says. 'So we had a bucket that we had to empty out twice a day because there were so many people. We had to pack like thirty gallons of water every day. It was *crazy* to do laundry. Every day, by hand, with a metal washboard.'

As we slice cheese, she tells me that she can't open her mouth too wide because her jaw was broken in a previous relationship. She tells me how her father had his leg cut off by a machete after a drunken argument. She tells me about her brother's suicide, and how it was she who found him. I look at her as she potters through the kitchen. That one person can hold all this within them. On the work surface in the kitchen is a framed picture of a drawing of a newborn and the epigram, *A baby is God's way of saying the world should go on*.

'Do you want to see something weird?' Matt asks me after dinner.

I sit behind him on the quad and we bump down a rutted lane that runs out of town away from the river, ducking branches that sweep across the track. When we have long since left any cabins behind he turns off beside a sign that says 'Fort Yukon Golf Course'. It is a boggy, half-cleared meadow, gently undulating. Clumps of alder dot the fairways, their trunks gleaming in the low sun like masts in a harbour.

'I have never,' Matt says, 'seen anyone out here.'

We take a walk around. Three flags mark the three greens. There is a pile of moose crap beside the final tee. He sits side-saddle on the quad and smokes a cigarette. He tells me that his dad drowned in the Yukon, just a few months after they moved here. He says that he likes to come out here to think.

All over town I am hearing that the kings are getting closer. Everyone has family in the villages downriver, so that they are able to chart the fishes' progress, and when I check my email I have the updates from Fish and Game. Rumours abound. There is a feeling in the air, a quickness in people's step. In front yards, in the hot, dry sun, people are mending nets and tinkering with outboards. I come across one man outside his house who is working on what appears to be some primitive fairground ride.

He tells me it's a fish wheel. I watch as he strips the bark from poles of spruce he cut that morning which are still gleaming with sap. He is building the whole piece from memory, cutting the joints with a hand axe, lining everything up by eye.

'Fish don't care if it's straight,' he says.

He explains that the mess of channels and islands on the Flats can make catching fish complex. There are few good eddies, the best places to set a net, and the maze of the river means that the fish could come from anywhere. Increasingly, those channels are now shifting, year on year, sometimes within a season, as rains worsen, permafrost thaws, and erosion increases. A well-placed fish wheel can mitigate some of these uncertainties.

Fish wheels appear to have grown out of the landscape, log jams with the life breathed into them. The wheel's origin is uncertain: maybe China, maybe France, maybe one of those ideas so pleasing and so simple in conception that it occurred independently to several fishermen, worldwide, simultaneously, as they sat and contemplated current. Certainly, though, it was introduced into America, rather than growing, unbidden, from the landscape. Probably first to the east coast, before drifting west, arriving in Alaska along with the Gold Rush. Driftwood logs, silt smoothed, form a raft upon which the whole contraption sits. Around a central axle, traditionally greased with bear fat, are two baskets formed from a lattice of spruce poles and strung with chicken wire. Crosswise to these, at ninety degrees, are two paddles, so that when the machine is set in the current, the whole thing revolves in the manner of a water wheel, the paddles catching the flow. It turns with lazy weight, seeming close to stopping on every revolution as each paddle reaches the apex of its turn, that limbic moment like the hang at the top of a roller coaster, before the momentum overrides the inertia and carries the wheel around again.

And then the calm explodes with sudden violence as a basket

scoops a salmon from the river and hauls it skyward, the fish swimming against air, battling this harsh new element and trying to find some purchase. In its struggle it drives itself across the basket with a series of thwacks, arching itself and letting go, bow and arrow both, and it is apparent that the fish's fight is part of the wheel's design, that it must propel itself towards the axis, and were it not for its struggle against the machine that has arrested its trajectory then it would simply continue to revolve, round and around, forever. But it does not. It reaches the wooden chute at the centre of the basket, and hammers itself down it into the waiting plastic tote, where it beats around the box until it knocks itself bloody and senseless. An old coffee can, precisely angled, provides a constant trickle of water over the fish, keeping them cool until someone turns up to collect them.

It is a lot of work to build one, and they last only a few years, but the man explains that once it is in the river and turning then it looks after itself. No need for all the messing about with nets. They are highly effective, and as such have even greater restrictions put on them than those in place for set nets. He hopes that Fish and Game will let him get it in the water soon. I ask him what he makes of all the closures. He shrugs.

'If I get my fish I'll be happy,' he says.

He tells me that if I've got questions about fishing, then I should go bother Richard Carroll, one of the most experienced hunters and trappers in town.

I find Richard's place not far back from the beach. I knock, and I am about to knock again when a voice calls from inside. Richard is lying on the couch with the curtains drawn and the television on. He looks up at me, a face like a boxer's. Athabascan, sixty years old. An *Alaska, Since 1959* T-shirt pulled tight over his thick and solid frame. He blinks at the daylight through the open doorframe.

'Well if you're coming in, come in,' he says. 'And close that door. Sit down. Are you one of those guys that come down and stink the river up?'

He gets to his feet, coughs and scratches. He pours a mug of coffee from the percolator and sets it down on the table in front of me.

'I guess you'll be wanting some food,' he says.

He pops a handful of pills and walks over to the fridge, squatting down to examine its depths. He is wound tight with energy, fizzing. There is some macaroni salad that he made for his girlfriend's birthday two days ago, but she never showed up to eat it.

'She went to Fairbanks to drink,' he says, by way of explanation. 'That was a month ago. She's been sleeping under a bridge. She's on the half twelve flight today. I gotta shave, I gotta clean myself up. Might get me some loving. Gonna need you to clear out pretty soon. Gonna get ugly. Now. Let's see what else we got.'

He opens up the freezer. It is crammed full of Ziplocs of chunks of meat, whole fish and bags of berries.

'I eat more wild meat than any other household in this town,' he says.

He dusts frost from labels, examining things, and then closes the door.

'I'll tell you what we can have,' he says.

He crosses to the sideboard where there is a large package wrapped in pages of the *Fairbanks Daily News-Miner.* He brings it over to the table where I am and he peels back the wet sheets. Inside there is a big slab of fillet with the skin still on, and it is unmistakably a salmon. But it is unlike any salmon I have ever seen. The flesh is a gutsy, intoxicating hue, so orange it is almost crimson; if something had swum thousands of miles to be here, battling storms and killer whales, this is just how it might look. The skin retains its vibrant polish, like a piece of bumper: it had many miles left to swim.

'Is that a king?' I ask.

Richard smiles. 'The first of the season.'

This is it! Here I am, in this house with a man I have just met, and he is offering me a piece of the first king of the season. One of Fort Yukon's young hunters had caught it last night, and he had presented it to Richard's mother.

'The real good hunters share,' Richard tells me as he takes up the cut of it his mum had given him and lays it on a chopping board. 'And when they need help they always get it. The bad hunters, they don't show nobody. They take it home and hide it. And now I'm sharing it with you.'

I feel so undeserving. What this means for me is superficial compared to what the kings' return means to those who live here, and I feel like the worst kind of tourist. Surely Richard should be sharing this with people who have just made it through eight months of winter, whose generations have been entwined with the salmon's generations for millennia, instead of with someone who just floated in and buys his salmon in the supermarket when he feels like it.

But then, there was always a certain astonishment amongst early Western visitors at Indigenous notions of sharing and cooperation, so at odds with the acquisitive cultures from which the explorers hailed. 'The Indians,' Columbus reported to the Court of Madrid upon his first arrival in the Bahamas, 'are so naïve and free with their possessions that no one who has not witnessed them would believe it. When you ask for something they have, they never say no. To the contrary, they offer to share with anyone.' He quickly realized this would allow him to obtain as much gold and as many slaves as the King and Queen of Spain might wish for. Charles Darwin, in his account of the voyage of the *Beagle*, lamented the 'nomadic habits' of the Yaghan people, writing that 'the perfect equality of all the inhabitants will for many years prevent their civilisation'. But for community-minded cultures with little interest in personal possessions or hierarchies, this way of thinking made absolute sense. The anthropologist Hugh Brody, who has carried out much of his research in the Arctic, describes how hunter-gatherer cultures store both food and knowledge by sharing them with each other. And even now, in the twenty-first century, when bank accounts and freezers allow the Athabascans to hoard as much as any other capitalist, and when knowledge can be hoarded online, there is still a deep sense of community and care.

'Money,' says Richard. 'What you gonna do with it? You'd just end up losing it all to your relatives. You can't eat it. The person who can provide is better valued than the local millionaire. You're guaranteed sex if you can provide.'

Richard sets some rice on the stove to boil and starts to slice the salmon into strips. He passes me one, speared on the point of his knife.

'Eat that,' he says.

As I chew, the flavours concentrate into something more like meat, a vibrant, earthy, firm, thick flavour. It coats my mouth in oil. I tell him that it is delicious. He grins, as though it is a personal complement. A good amount of his teeth are missing. He says he lost them fighting. He stands at the stove and tends the rice. He is one of those men who make the kitchen look like a workshop when they cook.

He has pulled some photo albums out for me to go through. Most are taken at his trapping cabin, many hours up the Porcupine River, which meets the Yukon here. He pokes at a picture of the snowbound cabin with a finger sheened with fish oil.

'I was up there nine weeks once and saw nobody,' he says. 'I was hallucinating ghosts come visit me.'

There are photos of salmon, and photos of him posing beside skinned caribou, the nude animals laid out side by side upon the snow.

'I can clean a caribou in twelve minutes,' he says. 'I don't waste my time.'

The rice boils over; he turns it down. He hasn't seen a caribou down this way since he was a boy, when herds crossed upriver from here. He has to spend so much money on gas to get up to his cabin on the Porcupine that he has to kill at least fifty just to make the journey worth it.

Fort Yukon would no longer exist, Richard says, if it wasn't for king salmon. The summer chum that make it up this far are rarely in good condition. The autumn chum don't keep well; by the time they arrive the air is so damp it is impossible to dry them. He tells me that thousands of years ago a volcanic eruption in the St Elias Mountains clogged all the southern salmon-spawning streams. Layers of ash two feet thick can still be seen in cut banks along the Alaska highway. Hungry people drifted north, looking for another source of food.

'Get it?' he says. 'King salmon brought us into this valley. If it wasn't for the salmon, we'd be chasing caribou up in the Porcupine Range.'

He brings the salmon and the rice to the table. There is wasabi and pickled ginger. He returns to the fridge and comes back with some mussels, and sets them down beside the rest.

'I didn't know you got mussels in the Yukon,' I say.

He looks at me like I'm an idiot.

'They're from Australia or some shit.'

He picks up a nori sheet.

'I can't remember which way round this son of a bitch goes, it's been so long,' he says.

He lays the nori sheet flat on the table, spoons on some rice, lays down some pieces of raw king. It is delicate work and it becomes him strangely. He rolls the whole thing up and cuts it into pieces. He smears a piece with enough wasabi to floor a horse and pops it in his mouth.

'Woo,' he growls. 'That's what I'm talking about.' He pounds the table with a fist, sweat beading on his temples.

'Little boy, come here!' he shouts.

From somewhere in the back of the house emerges a small boy, wide-eyed, with a shock of hair and a Spiderman sweater. Richard passes him a piece of sushi, and he stands there, chewing solemnly. Richard watches him intently.

'Looks like you grew last night, Martin,' he says. 'Or is it just your hair?'

Martin smiles at him shyly. Richard pushes the plate towards him.

'Have another bite, son!'

Martin shakes his head. 'My stomach,' he says.

'Hurt?'

'No, full.'

'From what?'

He points at the sushi. 'That, because it tastes so good.'

Richard watches as Martin walks back to his bedroom.

'Sweetest kid I ever met,' he says.

'Is he yours?' I ask. He looks too young.

'He's my grandson,' says Richard. 'His mother's in prison. So I'm looking after him now.'

We sit there, chomping on the sushi, swilling it back with coffee.

'I've fished my whole life,' he says. 'I built my first wheel at sixteen. My dad was US commissioner, one of eight jobs in the whole of Fort Yukon. My mum, when she wasn't drinking, she was too sick to do anything. So I fished for my whole family. I'm the only uneducated one. The only one that was ever interested in trapping and all that outdoor stuff. I learnt everything I know from my uncles and cousins. Work hard all day and expect nothing back. The kids in class were just kids to me. In high school, it was my job to go check the wheel before class. My two cousins checked it the rest of the day. I'd get up early summertime. It was about a half-hour round trip if I didn't goof around. I caught more fish than my cousins, because those fish run in the nightime. I hadn't realized I was shorting them until they told me. I thought gee, I better give them some fish.' He tails off, laughing at the memory.

The radio has been in the background this whole time. Something catches his attention now, and he stands up, finds it under junk and papers, and turns the volume up. The woman is talking about his mother, Eva Carroll, paying her respects to a venerated elder. The formalities taken care of, she starts in with the press release:

'This is an announcement by the Alaska Department of Fish and Game, for subdistricts 5-A, 5-B, and 5-C. Effective 6 p.m., Tuesday, June 27, fishing with selective gear will close, and immediately reopen with 6-inch or smaller mesh gillnets and fish wheels on the regulatory schedule of two 48-hour periods per week. 6 p.m. Tuesdays to 6 p.m. Thursdays to 6 p.m. Fridays to 6 p.m. Sundays. During fishing closures, fishermen may use gillnets of 4-inch or smaller mesh of 60 feet or shorter length to target non-salmon species. Chinook salmon harvested in 4-inch or smaller gillnets may be kept. Please keep a pen in your boat for the purposes of labelling your gear with your name and cell.'

The announcer has a New Zealander accent.

'A pin in your boat?' says Richard. 'What the hell is a pin? These people are meant to be educated.'

'Are you going to put your net in tonight?' I say to Richard.

He shakes his head. 'I'm not fishing any more,' he says.

I am taken aback. 'Why not?'

'It's not right,' he says. 'I bet I spent over ten years sitting on the subsistence council in this area. And when you do that you see what's happening.'

For many people in Alaska, the first time they knew there was a problem with their fish was when people from Teslin, the first village I had been to, eight hundred miles and a border away, started coming to their meetings. Every time the Yukon River Panel convened, representatives from Teslin would make the trip, and in winter that often meant a twelve-hour drive, one way. They turned up with a ghastly vision of a future with no fish, and a prophecy that, if left unheeded, would play out further and further down the river. They described themselves as the canary in the coalmine.

'I remember this meeting,' says Richard. 'I met these native people from Canada, and they all remind me of someone I know. They all look like me, talk like me. And they're honest. And we hear from the headwaters, people who fish the headwaters, who have voluntarily chosen not to fish. And we're hearing all these Alaskans down at the mouth complaining that they're not making enough money. I stood up. "I'm going to go and sit with the Teslin people," I said. I said to them: "I am ashamed."'

That afternoon, after Richard has got himself some loving, I walk back to his house. He has invited me to join him on the evening tour. The tour bus is an old school bus parked out front in a graveyard of buses, *Fort Yukon Transit System* picked out in fading paint along its side. Richard is in a clean T-shirt, freshly scrubbed and showered, with curls of wet hair flopping over his forehead; he looks like Stallone or something. I watch his mirrored shades in the rear-view mirror. Martin sits up front. Richard speaks to me on the intercom.

'Sit further back,' he says. 'You'll stink up the place. Remember, everyone else is paying.'

We drive out to the airstrip to meet the tourists from the plane. There are seven of them, three from New Zealand, the rest from the Lower 48. They emerge from the eight-seater, squinting and stretching.

'I'm sure glad you made the trip, ma'am,' says Richard, as he helps each of them up the steps onto the bus. 'Thanks for coming to our little town.'

Fort Yukon was on the tourist trail a century ago. Jack London set part of *White Fang* here. It is the only village on the Yukon north of the Arctic Circle, and once it was the easiest place for travellers to get a glimpse of the midnight sun. They came on steamers from Dawson City and stood around, waiting for a break in the clouds. Sunners, they were nicknamed. The locals have been finding the things that amuse tourists curious ever since. Nowadays most sunners drive the Dalton Highway north instead, but Fort Yukon, only an hour's flight from Fairbanks, still draws those seeking an authentic glimpse of Indigenous life. 'Dress for comfort, not to impress', advises the brochure. This is Richard's thirty-eighth season on the buses.

'Well good afternoon, folks,' he begins over the intercom, once everyone is settled. 'I'll be your guide for the duration of your time here. I was born in Fort Yukon, over there in the old missionary hospital. I like to tell people that I was the last planned birth in town.' He cackles and revs the engine. 'I'm the city mayor here. I'm also the local undertaker. So anything you might need, don't hesitate to ask.'

We pull out of the airport and turn back towards town. It is not long before we have to slow for a man who is weaving down the middle of the road. Richard carefully edges past him, two wheels up on the verge. The man stares up at us, bug-eyed and blinking, as though we are a product of his mind. He raises the hand with the bottle in it in salutation. And then, as though he has sprung a leak, he sinks gradually to his knees and sags onto one side.

'Old Johnson's having trouble making it home,' says Richard over the intercom. 'He'll make it though. Somebody'll pick him up.'

The passengers glance at each other. Someone, uncertainly, takes a photo. This wasn't in the brochure.

The speed limit is 15 mph. We crawl along, down Airport Street, up Third Avenue. Richard points out the health centre and the bingo hall, open three nights a week.

'There's the school on your right there,' he says. 'It stays open up to 60 below. At 50 below you can stay at home with an excuse. Unemployment here runs at 85 per cent. All the educated people left in the eighties and moved to Fairbanks. The training programmes took them out. We're what's left. That's Maria.'

He points out a young woman as we pass, walking along the roadside.

'She's got no business around here. She just graduated last summer. If she stays here another two or three summers she'll get into alcoholism.'

We drive along First Avenue, past the liquor store, which closes at fifty below. There is a collapsing shack on the waterfront, a man comatose on the stoop and a scrawl of graffiti that says 'NATIVE' in big, unsteady letters.

'There's one of the four hotels in town,' says Richard. 'I think that one's a three-star.'

He guffaws. One of the New Zealanders laughs, too, and then stops when he sees that no one else is.

'Was I funny?' Richard will ask me later. 'I thought I was funny. Perhaps I drunk too much beforehand.'

We drive on. It is a strange sort of tour, but refreshingly honest in comparison to the glitz and hidden truths of the tours I had been on in Dawson.

'Now these are our tribal government offices right here,' he says, gesturing out the window. 'We handle all the social issues locally here. We got a court system that handles everything except criminal law. Everything is done through our local tribe. This is all federally financed. The federal government recognize sovereignty but the state government will not. Twenty-six people work here, all local. My dad ran the city here in the 1970s. He hired this girl right out of high school. Well, the girl showed up, she was pretty excited. Next day, he never seen her. Next day, never seen her. Well, finally she came in on the third day, he asked her, "What happened to you?" She says, "I forgot I had a job."'

That gets a big laugh on the bus.

'Well, you know it's kinda difficult,' Richard says. 'When you come from that generation that's never worked a day-to-day job, it's a hard thing to get into. When you're fur trapping and fishing and fighting forest fires.'

On the outskirts of town are a baseball pitch and a basketball court, surrounded by bleachers. We pass the tank farm, which holds the petroleum that powers the village's generators through the winter. We pass the cell phone tower, and the water-treatment plant.

'And this here is the Independent Arctic Circle Baptist Church,' Richard says.

We have already driven past the Fort Yukon Assembly of God and St Stephen's Episcopal Church. On their murals, ravens take the place of doves. Signs outside say that everyone is welcome.

'They've only been here about twenty years,' he says of the Baptists. 'When they first got here they made the womenfolk wear dresses. They didn't have many takers. After three summers of mosquito bites and three winters of frostbite, they gave up on the dresses. They do a bit better now.'

'How many people live here?' someone pipes up.

'There's about five hundred in the village,' Richard says. 'About four hundred related to you, Martin.'

He pats his grandson's little leg. This is an exaggeration: only about a hundred and twenty-five inhabitants of Fort Yukon are descendants of Richard's grandfather.

James Carroll moved from Minnesota to Alaska in 1910, sucked in by promises of gold. By the time he realized that the easy pickings were long since picked, there was something deeper that had stuck. He settled in Fort Yukon and began trading in furs. At the time, Fort Yukon was the furthest western outpost of the Hudson Bay Trading Company, a corporation with a reach that covered a twelfth of the earth's surface. James married Fannie, a local girl, and they had twelve children. The couple lie side by side, now, in the family graveyard, a white picket fence around the headstones. Richard points it out as we pass. This is where he will be laid to rest one day. Here is all of one family's rootedness, it's founding myth laid bare. Maple leaves of bronze entwine their names. They are surrounded by their descendants. *Thomas 'Scotty' Carroll*; *Alice Amelia Carroll*; *Clifton Carroll*; *James Emil Carroll, Sgt. US Airforce*; *Margaret M Carroll, Beloved wife, mom and grandma, Died 5-5-2016* – the plastic flowers on her grave not yet faded. Carroll is like other surnames I will hear throughout the summer down the length of the Yukon: Honea, Peter, Demoski, Fancyboy, Williams. The river carries the echoes of the generations, a poem to a place. I have never felt history so tangible. People know where they belong here.

We pull up alongside a woman who is walking along the verge.

'Here's the next ex,' Richard says to the bus.

The doors swish open. She is tall, thin, squinting at the sun. He says something to her. She looks at him, says something back. Then she carries on down the road. The doors swish closed.

'Not too talkative,' he says over the intercom. 'She's been on a month bender. You can't blame her.'

We reach the beach. Richard parks and opens the hydraulic door so that his passengers can get off for their photo opportunities. Everyone has heard Fish and Game's announcement and the beach is absolutely buzzing. The light is lower, thickening out, but the day is just as warm. Skiffs are racing back and forth, or are aligned along the waterfront, Alwelds and Applebys and Wooldridges. Outboards cocked out of the water, Honda 90s and Yamaha 4-stroke 115s. Windshields made from ply and perspex, or ripped from trucks, are bolted onto prows. Blistering paintwork, seats exploding foam, barge poles for the shallows. Men stand in small groups on the beach, talking fish and mesh. Someone is fixing a fish wheel to the back of a truck with a complicated web of ratchets and hitches. We watch as they haul it across the dirt, the men milling about and shouting, laughing, offering advice. It is the work of men who have never had to rely on other men for anything, men with the belief that any difficulty can and must be overcome by themselves alone, that there is no other help forthcoming. The wheel shivers about. They edge it down towards the water. There is a great cheer as it floats.

Richard turns round to face us, one arm hung over the back of his seat.

'Fishing season,' he says, 'is *on*.'

It is July. I have passed beyond the Flats, Fort Yukon is some days behind, and the current picks up again. The geese have moved onto the islands where they will spend their flightless weeks as they go through their summer moult. Their goslings wobble about. It is the call of the Canadian goose that I will hear in London, months later, in the winter, and when I do, outside on my boat on the canal in the dusk, I will be transported here. They travel several thousand miles to spend their brief breeding season in the North, and by the time the weather turns they will be moving south again. Florida Keys, New Mexico, North Carolina. There are those who once said, by way of understanding their annual eight-month absence, that these geese spent their winters on the moon. And why not? Many animals would leave for unknown realms, and return on the yearly cycle: the swallow, the caribou, the salmon. They could be relied on as the land could be, as much as summer followed spring; human lives were shaped by and contingent upon their journeys. The Pacific salmon of North America's west coast and those that spawn in Japan and Russia had shared feeding grounds for many thousands of years before the people of these places were aware of each other's existence. Did the ancient Tlingit of Teslin, or the Ainu of Hokkaido, ever find a snagged hook or broken spear in a fish they landed, some unknown technology, and wonder about other people, other worlds?

Few of the great migrations remain. The European settlers shot sixty million buffalo as they moved west across the plains; there are fifteen thousand left in the wild today. Flocks of passenger pigeons once darkened the sky for days; the last died in 1900, hunted to extinction. Of the billion monarch butterflies that flew from Mexico to Canada each spring there are only a fraction that remain, a combination of habitat loss, pesticide use, parasites and climate change. For an animal to demand not only intact habitat, but that we afford it the vast swathes of space required for its migration, seems almost anachronistic on a planet as cramped and human-centric as our own.

In the eighth century the Venerable Bede described Britain's rivers as having 'the greatest plenty of salmon', and a thousand years later, on a visit to the Highlands, Daniel Defoe could still write of 'salmon in such plenty as is scarce credible'. In Ireland they were so abundant that they were hunted with dogs. Regulations to protect their numbers have been around for close to a millennium; the geologist David Montgomery suggests in *The King of Fish* that salmon fishing may be the oldest ongoing regulated profession in the English-speaking world. In 1030, King Malcolm II of Scotland established a closed season on salmon fishing between Assumption Day (15 August) and Martinmas Day (11 November), during the months when they were spawning. In 1215 the Magna Carta decreed that the king's fish weirs be dismantled to preserve salmon for the public good. A fine levied on Irish tanners washing leather near to spawning grounds, imposed in 1466, is perhaps the first piece of anti-pollution legislation in the world. Such conservative measures persisted down the centuries, and violations were punished with heavy fines, months in prison, even death. Yet by the 1700s, salmon populations in Europe were wavering. Deforestation was affecting the flow of rivers, removing the many overflows and sloughs that were necessary for their spawning. The waste from breweries, slaughterhouses, distilleries, farms, tanneries and textiles mills, against a background of overfishing, hit the runs hard. And then came the Industrial Revolution.

Writing in 1861 for the periodical *All the Year Round*, Charles Dickens lamented:

> The cry of 'Salmon in Danger!' is now resounding throughout the length and breadth of the land. A few years, a little more over-population, a few more tons of factory poisons, a few fresh poaching devices . . . and the salmon will be gone . . . And are we, active, healthy Englishmen in heart and soul, full of veneration for our ancestors, and thoughtful for the yet unborn . . . Shall we not step in between wanton destruction . . . and so ward off the obloquy which will be attached to our age when the historian of 1961 will be forced to record that: 'The inhabitants of the last century destroyed the salmon.'

Early in the nineteenth century, Dickens writes, the fish sold at Billingsgate Market had been caught a few miles from London Bridge – three thousand salmon every year – but within twenty-five years the Thames was empty. The last record of a salmon caught in London is for 1833. Today it is extinct from most European rivers.

America looked on, but did not learn. Deforestation, canalization, industrialization; cotton mills and steelworks; armies on the march devoured whole rivers' worth of fish. Later came the pesticides and dams. On the east coast, salmon once reached as far south as New York; now, just one per cent of Atlantic salmon spawning on the continent use the rivers of the United States; the rest are bound for Canada. And Canada is not innocent: since the nineteenth century its clearcuts and sawmills have been catastrophically damaging for the fish. The journalist Michael Wigan writes of how in the Bay of Chaleurs, between Quebec and New Brunswick, the catch might once have exceeded a million fish in a single year. 'That conjures up a picture of Atlantic salmon,' he writes, 'nearly as numerous as mackerel today.'

In 1804 the Lewis and Clark expedition became the first

American outfit to reach the western United States. Canoeing down the Columbia River, William Clark recorded that the water was 'crouded with salmon' and that the number of spawned-out salmon 'on the Shores & floating in the river is incredible to Say'. It has been estimated that the Chinook run was fifteen million fish. The entire way of life of the Chinook Indians was predicated on this annual influx of protein, a civilization possible thanks to a dependable resource, and a fully fledged salmon economy that facilitated both trade and subsistence. Eighty-five years later, when Rudyard Kipling took a steamer down the Columbia ('the river that brings the salmon that goes into the tin that is emptied into the dish when the extra guest arrives in India,' he wrote), he saw the fish wheels of the canneries each taking hundreds of Chinook every night. And then, in 1937, the Grand Coulee Dam went up. Salmon stocks declined by 92 per cent, or fourteen million fish. In 1900 forty-five million salmon swam up the rivers of Washington and Oregon and California. Today that number is two million.

Abundance is its own enemy. I paddle for hours, for days, through an infinity of spruce, and it seems almost hysterical to think that a little logging could ever make a dent here. But Britain was once as forested as this. Despite a culture formed upon a land of seemingly infinite abundance, Indigenous beliefs eschew waste as immoral. It is not a question of scarcity, it is a question of respect. All through the Pacific Northwest and Alaska there are different tellings of the story of Salmon Boy, or the Boy who Stayed with Fish, or Shin-quo-klah, who drowned and was taken to live with the Salmon People after he threw a perfectly good meal of fish onto the ground. Under the ocean, the Salmon People taught the boy the proper ways to treat them: that if he was hungry he could catch one of them, but that he must return every single bone back to the stream, that nothing should be wasted. He lived with them for several seasons, until one spring the Chief of the Salmon told his people to prepare for a long trip. Hundreds of them came out of their houses and boarded long dugout canoes. The boy travelled with them, back to the land of men, and at each river the Chief assigned one canoe to enter it, except for the occasional creek they did not enter, because those were places where the people had not respected the salmon taboos. Shin-quo-klah returned to his own village, and his mother caught him, whereupon he transformed back into a man. He became a great shaman, and from then on he taught the village how to respect and protect the salmon to ensure they would always return. One spring, a long time later, at the end of his life, Shin-quo-klah saw a transparent salmon unlike any he'd ever seen before, coming up through the river towards him. He speared it, and in the same moment he died himself; that salmon was his soul. The people of his village laid his body on a raft and pushed him out into the water, and he floated down the rivers to the ocean, where he sank back beneath the waves.

Up ahead there is the bridge. It looks preposterous; even from miles away it already draws all focus from the landscape. It spans the width of the Yukon, seven hundred metres from bank to bank. The last bridge was at Carmacks, 750 miles before. There will not be another one for the next 800 miles to the sea. Cliff swallows have built their nests in the eaves and they tumble about the girders. Skiffs buzz back and forward, and there are many more up on the beach. The E. L. Patton Yukon River Bridge is 130 miles north of Fairbanks, and for the surrounding villages on the Yukon this highway is the cheapest way to town. Now that the fish are here the whole river has come alive. Many natives work in Fairbanks these days, but they will drop everything and make the journey – three hours in a truck, a few hours in a boat – at the rumour of an opener. Weekend aboriginals, the villagers call them, still enacting their traditional journeys to their fishing spots. Men are shaking off the city, backing trailers down the slipway, helping their families into boats. People come over for a chat as I moor up. The grapevine still functions on the Yukon, and they want to know who's fishing, how they're doing, whose smokehouses are full. There is more activity than I have seen in some time. It has the feel of arriving at a crossroads on some ancient spice route.

Crossing the bridge beside the river is the Trans-Alaska Pipeline. Some 799 miles long, 48 inches in diameter, the pipe bisects the state from north to south, forming a crucifix with the Yukon's flow from east to west. The crude oil travels the pipe's length at a brisk walking pace, traversing three mountain ranges, and arrives at Valdez on the shore of Prince William Sound twelve days after it set out. From there great tankers ship it south, to Seattle, to California, to Asia. The pipeline was built in 1974. The United States was in recession, and tens of thousands of engineers and drifters flooded into Fairbanks in the largest influx the city had seen since the days of the Gold Rush. The telephone exchange ran out of numbers. The Wild West reared its head again. A man I once met in a bar in Fairbanks told me he still has a photo pinned above his desk cut from the front page of the *Daily News-Miner*, throngs of pipeline workers and prostitutes comatose outside the bars of Second Avenue at noon. 'I never saw a photo like it since,' he said.

There are a number of watershed moments in Alaska's history. On Boxing Day 1967 oil was struck on Alaska's North Slope. America had bought Alaska from Russia in 1867, a deal negotiated by William Seward, then US Secretary of State. It was a purchase much derided at the time: the press dubbed it 'Seward's folly', and the new acquisition as Walrussia. Not sure what to do with their new half billion acres, it was governed as a far-flung territory, with all the lawlessness that entailed. Statehood would not come until 1959, with the United States capitalizing on Alaska's strategic military importance vis-à-vis Japan and Russia. But it was in 1967 that Seward's folly hit pay dirt. The oilfield discovered on the North Slope would prove to be the largest in the United States. Sixteen billion barrels have so far been extracted from Prudhoe Bay.

Once the Yukon was Alaska's principal route of trade, but it is the pipeline now. The state's economy and self-image are deeply wedded to petroleum. 'In Alaska, we live on the commons,' Governor Wally Hickel said in 2009. The industry provides one-third of Alaska's jobs and more than half its revenue. Alaska is the only state with neither income nor sales tax. Instead every Alaskan receives the Permanent Fund Dividend (PFD), a portion of the interest gleaned from the revenue generated from the oil. In 2015 that dividend hit a peak of over $2,000 per person. The only requirement to qualify for the PFD is to have lived in the state for a year; that alone tempts many people north, and gives most Alaskans a very concrete interest in the health of their oil industry. Even the most ardent environmentalist finds it hard to turn their back on their annual free family holiday. 'There's two things in the world I hate,' a man once confided to me in a bar. 'Snakes and taxes. Alaska is the best place on earth.'

In 1988, at the pipeline's peak, it transported two million barrels daily. Now it is half a million. With Prudhoe Bay's production in decline, the crude's brisk walking pace is slowing to an amble, and in the brutal Alaskan winters that means wax buildup inside the pipe. By 2025 the daily throughput is forecast to drop below 300,000 barrels, at which point it will be unviable. It is in the process of becoming, as former Lieutenant Governor Mead Treadwell put it to me, the world's longest chapstick. As revenue dries up Alaska is running a deficit of four billion dollars, and new sources of petroleum are far from guaranteed. The expense of opening new reserves in such remote and harsh terrain, the fracking boom in the Lower 48, the falling price of a barrel of oil, the eventual phasing out of fossil fuels – all are putting the state's finances in an increasingly untenable position.

And as the deficit deepens, another expense is looming. In 2009 the US Army Corps of Engineers assessed the impacts of erosion in Alaska, erosion exacerbated by climate change as the permafrost degrades and the severity of storms increases. Twenty-six Indigenous villages were listed as 'priority action communities' for which 'immediate action' should be taken, and many of these villages are considering relocating. Moving a village, with all its associated buildings and infrastructure, comes in at around $500 million a pop. Another 152 communities were determined to be suffering from erosion of a severity not yet requiring that steps be taken. There are only 213 native villages in Alaska. Highways are buckling, foundations caving in. Trees have slipped to such rakish angles that they are known as drunken forests. It is tempting to see the absurdity in a state more affected by climate change than any other attempting to generate the majority of its funds using the very resources that are fuelling it. But beggars can't be choosers. Or as Lieutenant Governor Treadwell put it to me, leaning forward across his desk, 'My personal advice to you is don't try and sound brilliant by saying that we're a paradox. We're not a paradox. We all live in the same world. Oil pays our bills. Okay?'

I walk up the slipway from the river to the road. In the parking lot is a diner, the Yukon River Camp. I go in for coffee and a burger. This is the haul road up to Prudhoe Bay, and the emphatically named town of Deadhorse. If you ever watched *Ice Road Truckers*, this is that road. Great rigs are parked outside, bringing gear up to the oilfields, and alongside them is an assortment of camper vans and bicycles and Harleys. This is the last place to get a burger for quite some time, whether going south-north on the highway or west-east on the river. Except for the truckers, everyone here seems to be on some mad journey of their own devising: a drive from Texas to Alaska, a cycle from Patagonia to Alaska, a walk from Nova Scotia to Alaska, or simply some inexplicable need to straddle the Arctic Circle. Whatever road you might have started on, whatever your mission, at the end of this road is where you stop. Some 350 miles north, gazing out across the grey of the Arctic Ocean, your back to the oil rigs, wondering where next.

I camp one night in the woods beside the road. I am excited. Ulli Mattsson, my partner, is joining me here for the rest of the paddle to the sea. She turns up the next morning, dropped off by a friend from Fairbanks. She is fresh from a transatlantic flight and pale from an English summer, but she could not get away until now. We have known each other for three years, but it is the first time we have done anything like this together. We have scarcely been in touch since I started out six weeks ago, but it will not take her long to settle in. Ulli is from Boden, Sweden, way up on the Arctic Circle. One of her ancestors was Sami, the original reindeer herders, and this landscape is hers. The boreal forest sweeps through Boden, too. And although she left Sweden almost twenty years ago, she will find instant recognition in the food, the trees, the animals, and in the outlook of the people. It is wonderful to see her, something familiar in all this strangeness.

Civilization done with, we walk back to the river.

A series of creeks enter on the north bank of the river: Little Salt Creek, Sarah Creek, Susie Creek, Hannah Creek, falling out of the Rampart Mountains. These creeks are little scraps of paradise. Not for the first time, I think that turning off the Yukon feels like pulling off the highway onto quieter country roads. The creeks' banks are lush with grasses and with fireweed in full bloom. Butterflies flit about. There are the little birds that I rarely hear and never see on the river. Sheltered from the Yukon's constant headwind, the heat is sweltering. We take off our clothes and go brown. The creeks run clear, and beneath the surface I can see grayling, their peacock-eyed dorsals loose and feathery, gusting in the current like unreefed mainsails. We dive in. It is startlingly, achingly, wonderfully cold, and the sweat and grime and bug bites and sun cream evaporate in an instant. We whoop and shout, two animals in the landscape.

We are lying on a blanket, reading, in no hurry to get back in the canoe, half at peace, half wired for noises of bears, when we hear a plop in the river. Ulli starts, props herself up on her elbows. It comes again. By the time I rouse myself enough to look, Ulli is up and has the rod from the canoe and is standing in the shallows in her boots and not much else. I watch her cast and reel, cast and reel. Ever since I have known her, she has been unable to pass a body of water without fishing it or jumping in it, and invariably both. I have put this down to her being Swedish, or a Pisces. She has a ten-dollar rod I bought from a man in Fort Yukon, and within minutes she has a bite. It is a big fish. The little rod bends. She lets it run this way and that, and slowly, slowly, works it bankwards. She shouts for me to help. I jump into the shallows and manhandle the fish as she pulls, soft and viscous between my hands, and finally we have it on the bank.

It *is* a big fish. And as I am wondering how we are going to approach it, Ulli pulls out her knife, quick as a reflex, and drives it through the top of the creature's skull into its brain. It quivers and falls still. This is a side of Ulli that I have not seen. I am not sure if I am more stunned by her precision or by the strange creature lying at our feet. Its mouth is prolonged, like a pike. Its skin is olive with rough stripes of dull crimson, like the markings of a tiger. I think it might be a salmon but, embarrassingly, I have no idea. It looks nothing like the fillet that I saw on Richard's table. We are aware that we have just rung a dinner bell for bears, and hastily, she guts it. Unable to remember what to do with the offal in the many conflicting Indigenous stories I have read – bury it, burn it, return it to the river – we leave it on the bank for whatever might come along, and hurry back to the canoe with our prize. That night we barbecue it, a little lemon, a little salt. It is magnificent, juicy and tender, and we feel exceptionally proud of ourselves.

The following day is heavy with heat. A man stands in the river, fifty, wiry, in a singlet and jeans sodden up to his thighs. He hefts a fish with his knife stuck underneath its bottom jaw, and his whole body tenses with the weight of it, his thick muscles bunched, showing their veins, and the fish trailing slime like drool. He sets it, belly up, in the crib, a sort of sawhorse for the purpose. He severs the head, draws the tip of his knife along the belly from the anus and eviscerates it. Blood swills in the water round his boots. The gulls flock and dive about him. He bungs the whole mess of innards in a tote, where flies drone about the slop of roe and heads and guts, all of the innards except for the heart, which he sets on a flat rock to one side. Incomprehensibly, it twitches, first one half, and then the other. A pyramid of heart, the size of a walnut, with the electricity still running through it. It is a grotesque approximation of life, like some macabre Victorian experiment, quivering about on the slab. We all watch it. It is minutes until it stops. The man lifts the fish, his thumb in the cavity, his index crooked into a divot at what might be called the fish's shoulder; it is slippery as soap. He carries it up the riverbank and slides it onto the cutting table.

There are eleven kings remaining in the bottom of the aluminium skiff. Many are griddled with scars from the nets they have just fought against, and many are missing chunks of fin and flesh from the walrus and killer whales and lamprey and seals. The fish that were bound for the border and further into Canada are still a heady silver, whilst the kings that were nearing their spawning grounds have taken on a pink blush, as though embarrassed. The man selects another. Its back is rounded, and you can tell it is a male by how the head has lengthened to a snout; the teeth have grown, and the jaw has curved up and in upon itself to form what is called the kype. This is a mouth no longer adapted for feeding but for fighting, for keeping other males off at the redds. The new hump of its back makes it harder for a competing male to get in a bite across the body. Its stomach has shrunk, its sex organs have enlarged. Its fins have thickened and its skin toughened, absorbing its scales. Some species change so much – the sockeye from silver to arterial red, with a green head and a mouth like a crocodile – that for a long time they were believed to be separate species. Up to 50 per cent of a salmon's muscle and 90 per cent of its fat will be absorbed by the end of its migration; they are hard-wired to breed, everything else is sacrificial. This is why, the man explains, if you get them at the very end of their run, they taste like shit.

This is Joe Burgess' fish camp. Joe is standing in the shade at the cutting table in cargo pants sagging with slime, while the other people here are a mix of his family and his friends. Joe had called us over from the middle of the river and had pressed on us coffee and the remains of their breakfast. He had already heard that I was coming from his son who I had met upriver. Joe is perhaps in his sixties, with long wisps of grey beard and grey hair, a Stars and Stripes baseball cap. There is a dog collapsed in the shade at his feet, panting with the heat. Joe's knife is wickedly sharp, long and supple, and after every fish he sharpens it again. He runs it the length of the salmon, from tail to collar, lifts off the fillet, flips the fish, and slices the other side so that the backbone falls away. The flesh left behind on the spine is so thin that you can see the light through it. He trims away the belly, and then the dorsal fin, pushing down with all his weight on the back of the knife. Once people made boots and parkas from the skin, curing it like leather, so thick and tough it is.

Joe has a house upriver, but for a couple of months of summer he will pack up his life and live down here at camp. He is in the minority. Most people have jobs to hold down these days, and it is hard to justify taking a month off when Fish and Game might only give you one six-hour opener for fishing in a week. Often just one man will run a camp now, rushing off to check the set net on his lunch break. The price of gas, the price of time: it's the only way to make it work now. But Joe is retired, and for him time has lost its monetary value. The people down at camp with him are a collection of the poor and the misfits and the elderly from the villages around, those who can't get enough money together for a boat, or those who can't stop drinking long enough to fish.

'Everyone has a job here,' Joe tells me. 'The elder's job is to sit here and watch the river. The children's job is to take him coffee.'

Duane Aucoin, who I had met in Teslin at the beginning of this trip, had spoken to me about these wider functions of a fish camp.

'If we tried doing a healing ceremony,' he said, 'some people could say: "Oh, I don't want to be part of that. I don't need healing." But when you have a salmon camp, when you have the welcoming home ceremony, when you have the songs and the dances and the feasting, that's not a threat to people who might not normally want to go there. That's something that's safe for them to participate in, and then they find healing, and they don't even realize it.'

Jessica, Joe's wife, across the table, takes the fillets from him one by one, and I watch this dance of husband and wife, the steps perfected over years. Both of them cough and hawk spit as they work: there's a cold been going through camp. Jessica cuts a strip lengthwise from each fillet with her knife, a traditional knife in the shape of a half moon, tempered steel with a handle made of antler on the flat. The Yup'ik call it an *ulu* ; it is used by different tribes throughout Alaska, but only by the women. The big fillets are still the length and twice the width of her forearm, and she scores the flesh with a series of diagonal cuts an inch or so apart, so that when she tugs at each end of the meat it divides into a series of chunks, a gap between each one, held together by the skin. In this way when they hang the air can circulate, and as they dry they will not spoil. The flesh is the colour of blood oranges.

In an empty tote, Joe mixes creek water and salt in quantities that appear haphazard but that he assures me are not. As with much traditional food, the ingredients are starkly simple, the permutations endless. Some use less salt, some use more. Some use some basic spices. Some brine for ten minutes or twelve, or eight. Joe brines for twenty; that is my job, to check the watch. Some air dry before smoking, some go directly to the smoker. Some smoke with cottonwood, some smoke with alder. Some smoke with green wood, some with seasoned. Such refinements can be endlessly debated. Every family does it differently. Everyone I meet thinks that their mum does it best. Joe piles handfuls of salmon into the brine and then scrapes the cutting board clean of fins and mucous and waits for the next fish. The man lifts it, the largest, perhaps a thirty pounder, to the crib, and the whole contraption falls to bits.

It takes another hour to process them all. When they are done, I see that lying at the bottom of the skiff is another fish, and I recognize immediately its green hue and red stripes as the one that Ulli caught. Having seen the others, it is evidently a salmon, but something altogether different.

'What's that?' I say.

Joe steps back from the table and stretches, with his hands at the base of his back.

'That thing?' he says, pulling a face. 'That's a chum.'

'Don't you want it?'

'I wouldn't feed that to my dogs,' he says.

Jessica unlaces her apron. 'I'm gonna go lie down,' she says.

'I'm gonna go cook these hearts up,' Joe says.

Ulli and I follow him up the bank towards the fire. He says we should go look at the smokehouse. I pull back the tarp from its entrance and step inside. It is cool, out of the sun. It is a wood-framed structure of poles and uprights, covered over with tarp and corrugate. Smoke rises slowly from the smouldering alder; light glancing through the cracks and gaps catches the thick smoke in its beams. The noises of the camp seem very far away. It has the air of a cathedral, a place for the worship of fish. The salmon are hung in strips and fillets, slung over the spruce poles like stockings left out to air. In the rich dim light, they glow. They are starting to crust and oil beads at the bottom of the strips. The fish will spend up to two weeks in here with the smoke on them, the fire constantly tended. After that, stored in Ziploc bags, they should keep until the next time the salmon run.

Joe is sitting by the embers of the breakfast fire, stoking it up. A couple of old men are also there, nursing mugs of coffee. He puts a skillet on the grill and pours in a slug of oil. He scrapes in the salmon hearts and before long they are sizzling. An old lady summons Ulli to the other side of the fire. The woman has a bucket full of bellies, and she shows Ulli how to squeeze the blood from them, before submerging them in brine and weighting it all down with towels and stones. I watch them working, side by side. Joe passes me the skillet of hearts.

'Eat your heart out,' he grins.

I take one on a fork. It is the first of several bits of fish that I will learn you can put in your mouth. Later, downriver, I will watch as heads are split in half, cleaved between the eyes, like two-dimensional versions of themselves, and then wrapped in tinfoil and baked in coals. You can eat the piece of fat behind the eye, the cheeks, the bits of cartilage in the head, and I pick out thin slivers of meat and fat from the inside of the face, like cleaning up the carcass of a chicken. I like the taste, it is dark and rich, but I don't think I could eat much. There is a different quality to eating it, as though this is about more than food. As though I was transgressing some cultural taboo that was more deeply seated in me than I'd realized. A strong tenet of Indigenous belief is that no part of the animal is wasted, although today people are less scrupulous than were their ancestors. The kids don't like the hearts. Once the offal would have become bait for the trapline. The fins can be fried up like crisps.

Nina, Joe's daughter, emerges down the steps from camp, blinking at the sunlight.

'The crack of noon,' says Joe.

She rubs at her eyes and smiles. She lives in Fairbanks, and it is the first time she has been out here since she was a kid. She is wearing the T-shirt that she slept in, a grizzly bear with a surfboard on its back.

'Go and hang those strips,' says Joe. 'Make sure you wash the poles down first. Then you can have some breakfast. You come to camp, you gotta work.'

Many of the kids have left for the cities now, Anchorage and Fairbanks. Many of them migrate back with the salmon, but many don't. The city offers different pulls to the traditional ones here. A week or two later, in Ruby, another couple of hundred miles downriver, I will meet Isky Sartin. He comes down to the river where we are camped one evening to give us a stack of firewood and some fish strips that his dad has just made. He is in shorts down to his calves, wrap-around mirrored shades, a buzzcut. His mum is from New Mexico, he says, a Tewa-speaking Pueblo Native American. His dad is from Arizona, looks like Tom Petty, and was the first man to grow corn north of the Arctic Circle. Isky moved to Ruby in his mother's womb; he will soon be twenty-six.

A few years before, he had left town and gone to Vegas for a girl he met online. They had two boys together.

'Sin City,' he grins. 'I'd rap for money, me and this other dude. Talking shit to people until they bought us beer. My homey was sixteen. We were just being a couple of hoodlums.'

He found work as a doorman for a club, raking in the money: free drinks, free chips, good times. He had mobster connections in the city, he says, back from when his granddad was a used-car salesman there. His mum would post him salmon strips down, and he'd eat them up in secret when his girlfriend was out because she couldn't stand the smell. They broke up. He moved back just last week. He misses his boys. He thinks she just might move up here if he got himself established. There are a couple of women shouting, laughing, somewhere across the town. A chainsaw starts up in the woods. From the house at the end, beside the river, someone is wailing out 'House of the Rising Sun' on a tortured guitar.

'I missed how loud people were up here,' he says.

I have picked raspberries from the riverbanks and we share them between the three of us. A pike that Ulli caught is roasting on the fire. I ask Isky what he'll do now that he's back. There is money to be made here, he says, if you know how to go about it.

'I could make these young guys work and provide the money,' he says. 'You know. Let's go get two hundred gallons of blueberries and make some money and go to Fairbanks and party and do it *right*. Dollars don't even pass around four times here. We need to look after our own.'

He fills his pipe with weed and lights it, and then looks out at the river.

'The thing you have to understand,' he says, 'is we gotta make it a cool thing again to go hunting. We got to view subsistence as a term of endearment, rather than something that makes us primitive. Have you met the Williams brothers? They still go out and live self-sufficient. They hunt and trap. And they got their own homes and the prettiest girls in town. They're just *ballers* at it.'

'But young people are fishing, aren't they?' I say.

He shrugs, and points at the river. There are a couple of boats out there on calm waters.

'Not to the extent of being self-sufficient,' he says. 'There's no reason to live here if you're not doing that. It's part of their spirituality without them even knowing it. Before it wasn't such an option to fuck off and buy stuff from the store. Now you do your berry picking and get a bottle and that's it. All they want to do is drink and make babies.'

We watch the river passing for a while. Isky exhales slowly.

'If people don't at least take on the work of their fathers then this town will die,' he says.

Isky's father, Ed, runs a small farm over on the far side of town. There is an apple tree growing in the centre of the polytunnel. Cabbages, lettuce, chilli peppers, cilantro, broccoli, beans, zucchini, tomatoes. A Siberian strain of garlic, a Siberian strain of pot. Comfrey and poppies for medicine. Outside there are hives, and two tethered baby goats. Chickens muddle around, inside the house and out. I have seen some gardens down the river, but infrequently, and not like this. It is not because of climate. Twenty-three hours of sun and untouched, alluvial soil means that Alaska's cabbages break world records and the marijuana, as Isky tells me, takes prizes in the International Cannabis

Cup. But anything fresh sold in the stores is flown in, and the vegetables are limp and expensive. After 9/11, when every plane across America was grounded, the shelves practically emptied. Alaska has the highest rates of heart disease and diabetes in the United States. Growing food makes good sense here.

'What's hitting these people is this Crisco and white sugar and white flour and the denatured cornmeal and the canned goods,' Isky's dad says. 'Have you *seen* Crisco? I use it to polish my boots.'

'My daddy said that God's a farmer,' Isky says. But not everyone agrees. For Alaskan natives, the farmers arrived with the Christian God, about a century ago. It is a fundamentally different way of thinking about the world.

'Why don't you grow some vegetables?' I once asked a native man in Newtok who was complaining about a bad moose season.

'We're hunters,' he said. 'It's not what we do.'

Yet some native villages are now beginning to consider raising reindeer and bison, and cultivating berries, as a changing climate makes resources less predictable. Back at his camp, I ask Joe how the fishing's been.

'Not great,' he says. 'But a bit better than the year before. That's kind of what we go on now. Is it better or worse than last year? The water's too high. The highest I've seen it this time of year.'

He points at a tree stump down the bank.

'One year at break-up the water came up to there. We had to move out. We had all of our possessions in boats and we took them up through the woods behind the house. And there was this log floating through the woods, and sat on the log was a rabbit and a fox. Honest to God. Just that far apart, not looking at each other.' He smiles. His bottom teeth are rotten out. 'Sometime you have to compromise,' he says.

But for now, it is fishing season, and people are thankful to be fishing. As we paddle away from Joe's camp, we see other camps that are occupied also, and boats out checking nets, and fish wheels turning in the flow like grandiose water features. It is nothing like it once was, so they say, but after weeks of an empty river, it feels crowded here. Ahead of us the banks draw in and rise, and for the first time in a long time I am paddling in shadow. The Yukon narrows. On the map this is marked as 'The Rapids'. Cliffs rise above the water until we are moving through a gorge. There are peregrine nests up in the crags, marked by white stains that are centuries of shit. We drift, enjoying the speed. With the paddles out of the water, the other sounds come clearer. The separate chatterings of current against each bank. The shouts of men at camp, hauling up the catch. Distant birds that crackle like something tropical, and ones more distant yet, a falcon's thin, far shriek. I lie back and listen harder. The hum of insects, the wind in the spruce. The rub of silt against the hull. Sliding beneath the water's opaque surface, on the inverse, there is an equally persuasive current of salmon, as urgent, as unrelenting, moving towards the river's source. I see nothing, but I know that they are there.

By way of precise definition, the flesh of the fish hanging on the racks in Joe's camp would score a 29 on the SalmoFan, which is equivalent to the printing-ink colour Pantone 1655-U. The SalmoFan resembles a swatch of paint colours ranging from a pale rose to tomato, from which producers of farmed salmon select the exact shade they require for their fish's flesh, and then add the requisite amounts of canthaxanthin and astaxanthin to the feed. Farmed salmon, lacking the krill and other crustaceans that naturally dye wild salmon meat, will otherwise turn out a somewhat unpalatable grey. Some 66 per cent of consumers, reportedly, favour colour 33 (one shade below tomato). The Dutch multinational DSM, who produce the colour fans and dyes, claim that dark-coloured salmon commands a premium of up to a dollar more a pound compared to lighter-coloured flesh.

'Light-colored salmon attracts a different consumer demographic,' they write, 'being more frequently purchased by minority groups than dark-colored salmon.' But the most effective way to encourage spending is to present shoppers with a choice: 'DSM's research also shows that volume sales of salmon are increased when consumers are offered the choice of either dark- or light-colored salmon. Providing both flesh colors therefore stimulates overall sales.'

Up until the 1960s, the salmon was an exclusively wild fish. Today, 70 per cent of it is farmed. The salmon industry is valued at ten billion dollars worldwide, more than any other fish. It is the UK's biggest food export. In Scotland, the salmon market is worth more than its lamb and beef markets combined. Bucking a trend, Alaska chose to ban fish farming in 1990, to protect both its wild fish and wild fishermen. Rather than compete on price, they chose to promote the value of wild over farmed and to sell it at a premium. '*Wild salmon don't do drugs*' was a bumper sticker at the time.

Most consumers have little awareness, and little interest, about whether their salmon is wild or farmed. Most people assume it is probably wild because it's, you know, a salmon. Marketing uses bears and fjords and gruff men in sou'westers. In Great Britain, Marks and Spencer's salmon comes from Lochmuir, a loch that does not exist. Its name was chosen by a consumer group as having 'the most Scottish resonance', and the salmon under its trademark is farmed at five separate sites around the country. Salmon has maintained its image of being wild and healthy and delicious whilst morphing from a luxury product into one of the cheapest fish in the supermarket, the most popular fish in Britain, a staple of cut-price sushi and sandwiches and fishcakes and ready meals. But this democratization of food has come with hidden costs. We are not in the habit of farming carnivores, and to cage one that is more accustomed to migrating thousands of miles in a space no bigger than a bathtub has inevitable ramifications.

Salmon feel fear, and pain: the cages exacerbate both. They are prone to heart disease with such little space within which to swim. Their bodies are so altered through confinement that they are often referred to as a different species: *Salmo domesticus*. It takes between 2.5 and 5.5 kilograms of wild fish to produce one kilo of farmed salmon. Not only does this take food out of people's mouths – the anchovy market in Peru is now almost entirely sold as feed for salmon farms – it is also devastating marine ecosystems. Links have been drawn between the chemicals used in farming and carcinogens in people: for salmon imports from some countries, the US Environmental Protection Agency advises eating no more than six portions a year.

Sea lice (*Lepeophtheirus salmonis*) occur naturally in the wild, but in the crowded environment of a salmon farm they thrive. When wild salmon pass the fish farms the lice transfer to them, along with other diseases: ten lice can kill a smolt. 'There is a risk that salmon lice may kill off entire populations of wild salmonids,' writes Norway's Directorate for Nature Management. This has been going on for years, but it is the repercussions on farmed fish that is making the industry take notice.

The lesions that the lice cause, chewing through the skin and feasting on their mucous, can result in fatal infections. A combination of warming waters, which favour lice, and an increased resistance to antibiotics and pesticides, has meant a decline in global production. Scottish farmers have increased their use of chemicals tenfold over the past decade, with little impact but for the serious pollution of their lochs. The Scottish Environmental Protection Agency is now considering the construction of the world's biggest salmon farm, three times the size of the current largest. It could farm two million fish at once; it would produce as much waste as Glasgow.

Fish ranching was meant to mitigate some of these problems. Salmon are raised in hatcheries, transferred to salt-water pens as they grow, and then released into the ocean when they reach a size where they can fend for themselves. Ranched salmon can be labelled as 'wild-caught', a distinction missed by most consumers. Five billion juvenile ranched salmon are released into the North Pacific every year. People I meet along the river believe that this glut of fish is one factor in the Chinook's decline: five billion extra fish in the sea is a lot of mouths to feed.

As with all industrial-scale farming, there are worse and better ways of doing things. Allowing the fish more space alleviates some of the effects of cramped conditions. Land-based tanks allow for effective, if expensive, treatment of waste products. Keeping wrasse and lumpsuckers, small fish that eat the lice, in with the salmon, is an organic way of dealing with the parasite. But as with all farming, there is a tradeoff between sustainability, welfare, and price. Salmon is the cheapest fish on the shelf; supermarkets are inclined to keep it that way.

In 2017, the AquAdvantage salmon went on sale in Canadian supermarkets. Despite being the first genetically engineered meat to enter the supply chain anywhere in the world, it was not labelled as such. Canadian ministers have voted not to label genetically modified products in case it puts consumers off. The fish, produced by AquaBounty, is largely based on an Atlantic salmon, with a gene from an eel-like fish called a pout and a growth-hormone-regulating gene from a Chinook. The AquAdvantage can go from egg to market in half the time it takes a farmed Atlantic salmon. AquaBounty have called it 'the world's most sustainable salmon', which is a bit of a kick in the teeth for wild salmon, who had been managing perfectly well for several million years. Their fish require 25 per cent less feed than traditional farmed varieties, they say, and will bring jobs to the United States, although AquaBounty currently cultivate the eggs in

Canada and then fly them to Panama to raise them. Through an extremely convoluted, expensive, roundabout route, the United States might one day get back to producing as many salmon as its rivers did, naturally and effortlessly, a hundred years ago.

AquaBounty have so far only raised salmon on land. The company claim that their fish are sterile and that they will only be making females, so that the possibility of them escaping and reproducing in the wild is 'impossible'. Yet it is worth remembering the promises made when fish farms first came to British Columbia.

'We were told they wouldn't escape. They escaped,' said Jennifer Lash, director of the Living Oceans Society. 'We were told they wouldn't survive in the wild. They survived. We were told they wouldn't get upstream. They got upstream. We were told they wouldn't reproduce. They've reproduced.'

There are the economic arguments and the environmental arguments, and then there are different, deeply seated conceptions of the world. Subsistence and capitalism, hunting and farming, are two entirely different ways of being, but one has become so dominant that to question it now seems hopelessly naive. Yet after forty years of salmon farming along the coasts of British Columbia, First Nations have had enough. Recently, two fish farms have been occupied and have had eviction notices served on them by activists alarmed at the effects that these corporations are having on the wild salmon stocks, and citing ancestral rights to their land.

'You are renters,' Chief Willie Moon of the Musgamagw Dzawada'enuxw told the Norwegian-owned Marine Harvest. 'We are kicking you out and keeping the deposit.'

Some days he forgot to look. Some days, he thinks, as he sits on the stoop with his first coffee of the day, were like he was an insurance salesman or something, pounding the streets of New York or some God awful place like that. Some days he didn't notice the birds at all. Running through the day's to-do list: cut fish, dry fish, can fish, haul the drift, fix the winch, do the dogs. Though he'd been out here long enough, now, that perhaps his perspective of what constituted the rat race had somewhat warped. Yes, it was possible to imagine Stan Zuray in the city, his boxer's stance crushed into a cheap suit, his ponytail pulled back, broad shoulders, squashed nose, the Bostonian accent that belied his origins. It was possible, but it required an imaginative leap hard to maintain. To most people who knew him he made sense no place else; he had filled out into this country, expanding to fill all available space like the flow of the river itself. He rubs at his unshaven face, tugs at the braided necklace that he wears. He likes to joke he retired at twenty. The tip of his right index finger is missing, some accident with machine or chainsaw, that Alaskan rite of passage, the ritual circumcision.

It is a couple of weeks after solstice. He likes to sit up here in the mornings in the shade of the birches, looking down on the river unrolling. They call this spot The Rapids, although there are no real rapids to speak of, just a little riffle over Rock Island when the river is at the right height. The Yukon narrows so much here that in 1843, when Lieutenant Zagoskin came upriver exploring for the Russian Empire, he assumed he had made it to the headwaters and turned round and went back. Stan stands and walks into the cool of the cabin. The windows are open, the bugs have not been so bad this year. The room smells of spruce and coffee, and he pours another from the percolator, still muttering away on the hob. Beside it is a pot of moose soup that his son Joey knocked up yesterday before he and his mother went back downriver to Tanana. Take a chunk of moose meat, wild, local, organic, the whole nine yards, about the best meat you're likely to find anywhere on the planet, and mix it with white pasta and white rice and stock out of a packet. He couldn't understand it. He mixes up the batter for bannock, his own recipe: wholemeal flour, raisins, milk powder, a little bicarb, a little salt, and ladles the first dollop into a skillet. With his family gone, he can eat what he wants.

Everything back in its place, just how he likes it. The dishes done, notices pinned around the place about what recycling goes where. 'No metal lids, aluminum foil, diapers, glass in wood stove. Please.' Labelled plastic gallon jars: instant potato, pinto beans, popcorn, yeast. Beside the bed is a photograph, a printout of all the kids and grandkids, taken in Fairbanks. They are arranged around a flatscreen television, and on the television is Joey, the one absent sibling, but positioned just so, as though he is there with the rest of them. They are wearing grins of achievement, disbelief, the sorts of grins reserved for summiting mountains. It's from the premiere of *Yukon Men*, the TV show that he and Joey star in, the first time it went out. Stan flips the bannock, gives it a few more seconds and lifts it out, and pours the next into the pan.

Outside, the raven is up on the roof of the cutting house, whirring through its repertoire of electronic noises like one of those old modems. More calls than nearly any other animal on the planet, apparently, but it still didn't take long to feel like he'd pretty much heard all of them about a hundred times. He turns the radio on to block it out. There are two radio stations to choose from out here, the sport and the religious. The religious one carries the weather and flying conditions and fish openings. Maybe a tune, if he's lucky. It also carries the mukluk telegraph, all messages read out by the DJ in the same laconic drawl.

Rolly is headed out and has your medicine; he should be by your cabin by day after tomorrow.

Have Hank bring the supplies on his next trip, and add a slab of bacon.

May is recovering after having her appendix out. Doctor says she should be able to head home soon; next boat we think.

The screening of Cloudy with a Chance of Meatballs is cancelled tomorrow at the gymnasium. It will be rescheduled for a later date.

To Betty at Rampart: Delayed in Anchorage. Waiting for Cessna parts. Hope to be home by weekend. Will send another message Thursday night. Please feed the dogs. Rob.

To Mom in Tanana: Jenny's baby girl arrived Tuesday night; 7 lb. 2 oz. Both doing well. Plan to be home on Saturday. Bill.

To M. in Tanana. Bring toilet paper! – C.

He liked listening in, but it was nothing like it once was. Never mind the fish, it's the people out here going extinct. He could remember driving up the river at dusk, and all the kerosene lamps of all the camps coming on along the bank, and for each point of light, he would know who was there, how their fishing was going, whose smokehouse was full. There used to be a time when more or less every native on the river was fishing in the Rapids. There were killer eddies all over, and fish wheels all lined up beneath the bluffs. They've dug up all sorts of artefacts here, over the years. Knapped stone, fish bones, old cordage made of caribou tendon. Now it's just him and his family out here, and a few other old white guys.

Stan first came to the Yukon in 1970, age twenty. As a kid, he had run with the gangs in Boston, stealing cars and fighting, selling hubcaps and fake IDs. He wore a leather jacket with his hair greased back. He was stabbed twice before he was eighteen.

The corruption he had seen, in the cops, in the church, in his employers, had taught him a healthy distrust of authority that he would carry with him to the grave. He lived for a time in a commune on the outskirts of the city, but nothing was going anywhere. He wanted out. The sixties had peaked, the dream was dead, or dying. He sold his home-built hotrod for a VW and drifted west, to California, and then inexorably northwards. Oregon, Washington, British Columbia. There was a well-thumbed route to follow.

'The corruption of city life, the loneliness, and the realization that there was an alternative to it all, that sent me packing,' he says. 'I was part of the problem, and I didn't have to be.'

Casual work was abundant; land was cheap, or free. There was fishing, there was construction on the pipeline. And the north offered, too, a good chance of dodging the draft. Vietnam was cooking, and Stan lived 150 miles up an old mining road with ten legal Canadians and ten Americans on the run. It was the taste of community and back-to-the-land living he'd been seeking, or it was until one day the Royal Canadian Mounted Police turned up and rounded all the Americans up. All except for Stan, who had himself declared insane.

'Might as well order me a load of books and lock me up now,' he had said during his draft interview before he left the States. They didn't know what to make of him. Most of the people that tried to shirk the draft did something more persuasive – bad back, bad eyes, blood in the urine – or else they made a case as a conscientious objector. They passed him from one shrink to another, and finally the Military Entrance Processing Station categorized him a 4F.

'That's to say,' says Stan, 'I was nuts.'

Maybe he was, but he was honest. He was the youngest of his Boston crowd, and he'd already had friends returning from the front lines. They came back junkies. They came back crying, which was worse. He helped them as best he could. Late night, drunk, there were stuttered confessions of the things that they had done. Murders and rapes and things unnameable. Don't go, they said, don't go. And he didn't, although the guilt of it had never left him.

After the raid, Canada no longer felt good. He went back to Boston for his brother's wedding, with the idea of persuading his girlfriend to come with him to Alaska, but whilst he had been gone she had joined up with a Bible cult in Brazil. At the wedding he met Charlotte, a friend of hers, and she didn't have any plans. When he flew out of Boston, heading for Alaska, she came along for the ride. The Homestead Act of 1862 allowed anyone who had never taken up arms against the government to file an application for 160 acres of unappropriated federal land, which would become theirs if they lived and worked it for five years. Such a cavalier approach to land and property is hard to imagine now, yet homesteading was possible in parts of Alaska until 1986. By the 1970s legislation, directly linked to the discovery of oil in '68 in Prudhoe Bay, was beginning to tie land up, and Stan was one of the very last to get land in the interior.

Stan and Charlotte chartered a plane and unloaded everything they owned on the banks of the Tozitna. It was August, pushing winter. Quitting wasn't an option. They got a cabin up before the worst of the weather came, but they had no time to prepare for the winter, which is, after all, what summers are for. He shot a moose, and they ate the whole thing in two months. Heaped platters of moose steaks, three times a day. They gave a moose haunch to the dogs, who were starving worse than them, but they hoovered the whole thing up so fast they felt it was a waste. They ate the dogs instead.

They made it through. There was so much to learn. They had a baby, and it died. The next week they heard that a mother in Tanana was looking to give up her newborn for adoption, and they took it. They had people coming round the house to say how sorry they were, and while they were visiting they had others visit to offer congratulations. After eleven years Charlotte was ready for a change. For some people, the bush suits them for maybe five or ten years, until they prove to themselves that they can do it, and once they know that then it's finished for them. For others, people like Stan, it's the life that they were meant to be born into. Without recrimination, they went their separate ways. Years later he would marry Kathleen, an Athabascan from Fort Yukon, and start another family.

He drains his coffee and walks down to the beach, past the smokehouse, past the micro hydro, and he untethers his skiff and pushes out. The heat is dry and clean and good, the wind off the water. The hills sweeping up, the spruce falling to the river. The way the hills hem it in, it feels cut off and distant here, their own private corner of the Yukon. Stan draws an arc out into the river's flow and turns the prow upriver towards Steve O'Brien's camp. A diary Steve had showed to Stan, kept by a passenger riding one of the old sternwheelers a hundred years ago, tells of how they were forced to go ashore here for three days as they waited for a herd of caribou to cross, thousands upon thousands of them. Each appearing between spruce trees at the water's edge and stepping to the current, their nostrils and their withers above the surface as they strained for the far bank. All of them with a single common purpose, so much so that it appeared ritualistic, some collective rite with no high priest, conducted by the imperatives of their biology, bound for their calving grounds. They still came through when he fished here as a young man, but they had not passed this way in years. Someone shot a lone male up on one of the ridges a few years back, but that was unusual, even then.

The river is so high that Rock Island, out there in the middle of the bend, is completely submerged, just a whisper of current to show where it sits beneath the surface. Some years Stan would keep the pups out there, when they were too old to stick close to their mamma any more and still too young to tie up. They had forever caused havoc in the camp before he hit on the idea of marooning them in the middle of the river. He fed them every day or two. They didn't seem to mind. And it kept them safe from the bears.

Steve O'Brien's floatplane, a 1946 Aeronca Chief, bobs in the back current of the eddy, rubber duck yellow, a slap of colour, artificial and primary, against the usual colours of the landscape. The fish wheel he is working on is moored alongside, half-finished, just one of the uprights on. Steve must be up at his cabin. He lets Stan fish his eddy here, because Steve is only interested in the wheels. He builds a new one almost every year, each one improving on the mistakes of the last, striving for some quintessential form. Oftentimes these big trunks, the spars and the uprights, have drifted down from some tributary a thousand miles distant where the spruce grow tall and thick. Everything comes down the river eventually, given time and enough patience. It's just a shame, Stan thinks, that there's not enough coming up it.

It was Art who gave Steve O'Brien this camp. Steve had just split from his first wife, and he needed a new eddy to fish. Art was an old Athabascan they all knew, and Steve had taken up fishing with him when he had first come out this way. Art got drunk the day he married and didn't get sober for forty years, so he understood how marriages could be rocky. He sobered up in his sixties and his mind was still as sharp as a tack. When he was dying Steve asked if he could fish this place, and Art thought long and hard and then said yes.

It's a good eddy. Stan pulls his net. Four fish. Not too shoddy. He hefts one, thumb jammed in its mouth. Twelve pounds, he guesses. There was a time when each year they'd have a bet to see who caught the first fifty-pounder of the season. They caught one fifty-pounder between 2001 and 2013. This year Stan's average king was 7.3 pounds. As is his habit, he keeps a careful count.

The first time that he noticed something was different with the fish was during the mid-nineties when he was working in his smokehouse. When he built it he had put in the racks too close together, so that the strips above touched the ones below, which wasn't good for keeping the mould off them. One of those stupid mistakes. The fish on the lowest rack brushed his head when he walked through: he liked to tease the kids that that was how he kept his hair so shiny. And then one year they weren't touching any more, and he didn't have to duck.

Everyone had stories that had been passed down, but who believes fishermen when they talk about size? Yet Stan could *prove* that the kings were shrinking. Since 1996, he had helped with a federal study into the Chinook on the river. Stan had always loved numbers. At night, as a boy, while his brother lay across the room reading comics, he was under the covers with a torch going through his father's technical manuals. Piston sizes, torque strengths, calibrations and conversions. Stan hadn't been long assisting on the project before he started to detect some pretty serious holes in the data, if the aim was to determine not just numbers, but also the size of fish. The data sets were not long term, sampling was not always random, there was a mishmash of techniques. Stan didn't know much, but he did know that you couldn't trust people in authority. And that if you wanted a job done right, do it yourself.

Fish wheels had long been used by biologists to study salmon, holding the fish in a live box for later release. Yet a 2007 study by Fish and Wildlife into the effects of holding fall chum in a live box, and handling them, showed that those that spent time incarcerated had a reduced migration rate and were more likely to be recaptured further upriver, disoriented, injured, sometimes dying, factors that the study's authors put down to the stress that the fish had accrued. Stan liked solving problems – his father had impressed on him that there was nothing in this world that could not be fixed – and he set out to make the most fish-friendly fish wheel in existence. He replaced the monofilament netting of the baskets with rubber webbing, and he swapped the wooden chute for plastic. A system made from a solar panel, a computer, a video camera, a waterwheel generator, an infrared sensor, some LED lights, some car batteries and a wireless transmitter meant that the fish could be recorded on their journey down the slide, the images beamed to a laptop back at camp, so that there was no need to hold them in a live box. The fish slid straight back into the water, a little startled, maybe, but otherwise unharmed, and the salmon could be another thirty miles upriver by the time Stan came into the office the next morning and counted what he'd caught. Stan's design has since become the standard for fish-wheel projects across Alaska.

Stan found high-school kids from the villages along the river to help him with the project. They knew fish, and they knew boats. They called him a data Nazi: he threw the first three years of data out because it wasn't of sufficient quality. But slowly, learning as they went, they began to collect what fisheries biologists all up the west coast of America would come to recognize as being the best data set on the Yukon. And it was proving what the Alaska Department of Fish and Game were denying; that the fish were shrinking, the ratio of females to males was declining, and that a disease called Ichthyophonous was having an effect on stocks. People were asking ADF&G why their data wasn't as good as Stan's. Why he was better than ADF&G at predicting the number of fish that would make it across the border into Canada. For Stan, it was clear what had happened to the fish, because he had been there and he had seen it happen.

'People had probably been overfishing for a *long* time,' he says.

'You didn't need climate change. You didn't need all the ocean current theories. It's completely understandable.'

Stan remembers a meeting, just after the runs began collapsing, when he and Andy Bassich and a couple of others went to testify in front of the Board of Fisheries that something needed to be done. The Yukon River Delta Association had chartered a fifty-seater plane to bring commercial fishermen up to testify against them.

'Seventy-four people testified that there was no problem,' says Stan. 'They said if you cut down our fishing then we're going to starve. And ADF&G said no problem. They ruined it for themselves. That's the fucking thing. And I'm a commercial fisherman. It's too bad that they drove it into the ground that much.'

Stan sat on the board of the Yukon River Drainage Fisherman's Association, but they were heavily lobbied by the commercial fisheries downriver. Twice in 2007 he brought a vote on whether or not there was a problem with the king's size, and twice he was the only member of the board to vote in favour. In 2010, ADF&G suggested that if there was a problem with the king run, it might be because too many fish were making it upriver. Managers threatened Stan that his funding could be withdrawn if he didn't stop offering comments on his website about how the run could be improved. Fred Andersen, a former fisheries biologist, is more explicit still.

'I don't think they wanted the answer,' he has said. 'I really don't. I think that the very essence of this is that those guys just didn't want to deal with the shitstorm of protest that would blow up if they took the draconian measures that were required.'

In 2015, disgusted that the state still refused to use his figures, Stan gave up the work for good. He says it will take fifty, maybe a hundred years, to get the salmon size back up to what it was, and that's if the management is done right. There is at least one glimmer of hope. 'When humanity wipes itself out,' he says, 'then the fish will recover just fine.'

For now it's just him out here, and Steve, and the Campbells across the way. And all of them stiffer and greyer and more stupid than they used to be. Linda, Steve O Brien's first wife, doesn't even come any more, not since she got busted selling fish to a native couple in Fairbanks. It wasn't free to run a fish camp. You couldn't put salmon in your gas tank, and how were you meant to hold down a job if you spent your summers out here? Once the bureaucrats had turned a blind eye to it, but it wasn't like that any more. Linda never came back. Her camp sits across the water, collapsing back into the landscape.

Of course, that wasn't the whole story. Linda hadn't felt at home here since Eric died. When Steve and Linda split she married Eric, and they had lived on the opposite bank. They were all young and beautiful, back then. There are old photos of them, the yellows and the oranges intensified, astride the roof of a half-raised cabin, or skinning a moose, thick beards, topless, the faces of the men they were to become already there, peering out, but uncomplicated by any sort of future. And their babies, like their cabins, marks of themselves that they were leaving on this landscape, inhabiting the land, part of its rhythms and its logic. But such marks fade so much faster out here. The land leaves its marks on the people, rather than the other way around.

Eric was taken by the river, and it could have been any of them. One of those freak accidents. He was the most athletic of all of them, and he was the first to go. Arms like Hercules. But he'd hurt his back a few days before, and maybe that was why. He and Linda were out checking the fish wheel when their skiff came loose and started drifting with the current. No one wore lifejackets in those days. He dove in, but the river kept pulling the skiff out further, further, keeping it just out of reach, and he swam after it, like a child chasing a balloon. She took her eyes off him for just a moment as she walked the spar to shore, and when she looked back he had gone. There was so much silt in the river, he could have been inches under the surface and they still wouldn't have seen him. Just gone, out of their lives, in an instant. The silt fills your clothes, it fills your pockets and your boots. He floated up two weeks later, across from where the wheel was, pale and distended. It took two days to dig the permafrost deep enough to bury him.

Stan pays his net back into the water, and then turns his boat out and steers it out into the current, heading back downriver to his camp.

The bedrock of Rock Island, out there in the middle of the river, was once earmarked as the foundation for what John F. Kennedy declared would be 'the greatest dam in the free world'. The Rampart Dam, first conceived in the 1950s, would have created the largest artificial lake on the planet, a record it would still have held today. Bigger than Lake Erie. Bigger than Macedonia. Ten million cubic feet of concrete, 530 feet high and 4,700 feet across – almost a mile. The lake would have stretched upriver for 280 miles, flooding the Flats beyond Fort Yukon, and would have taken twenty years to fill, its waters leeching out across the watershed as though the very earth had sprung a leak.

The Alaskan senator Ernest Gruening had told the state legislature with authority that the land was ripe for inundation as it contained not more than ten flush toilets. This was almost certainly an overestimation.

'Search the whole world,' he continued, 'and it would be hard to find an equivalent area with so little to be lost through flooding.'

This was wilderness in the sense of God-forsaken, more akin to where Moses wandered than some untouched, untamed Eden. The lobby group North of the Range recommended that developers 'come forward with both guns blazing', which must have been a wonderfully liberating way to express yourself when compared to these days of cumbersome PR and environmental impact assessments. Yet a federal report found that: 'Nowhere in the history of water development in North America have the fish and wildlife losses anticipated to result from a single project been so overwhelming.' The 36,000 ponds of the Yukon Flats provided nesting grounds for over one and a half million ducks, as well as geese and cranes and waders. 'Did you ever see a duck drown?' the dam's proponents sniggered. There were nine villages upriver that were slated to be flooded, 1,200 villagers to be evacuated, and the lives of thousands more changed forever as the salmon stopped returning. Salmon can leap prodigiously, but not 530 feet. For several years they would have arrived at the foot of the dam, beating their brains against the concrete, and then they would stop coming. But hey, salmon don't drown either.

One study calculated that the dam would produce more than five times the quantity of power than the state of Alaska could consume. A legislator remarked, in private, that the dam would have served its purpose if it were blown up on the day it was completed, because in the end it was primarily a project conceived to channel federal money to the state. A brochure produced by Yukon Power for America (membership for schoolchildren at twenty-five cents a head) waxed breathless about its recreational potential: 'Freshwater boating and sailing . . . hunting lodges and fish camps on scenic shorelines . . . marinas, dock and float plane facilities – all accessible by rail, highway and air.' Of all the claims, this was perhaps the most delusional: that this vast new lake, hundreds of miles into the Alaskan bush, grazing the line of the Arctic Circle, would see families picnicking on its banks on the Labour Day weekend, and water-skiers carving arcs through its ice floes.

It was one of Alaska's formative environmental battles, one for cutting teeth on. For the first time in American history, a successful case was made for the protection of a river on the grounds of fish and other wildlife. A coalition of environmental groups stalled the project until 1968 when development interests turned their attentions to Prudhoe Bay for a quicker, less controversial buck. When President Jimmy Carter created the Yukon Flats National Wildlife Refuge in 1980 the dam was damned for good. How different The Rapids could have been. Stan, sitting in his cabin beneath five hundred feet of water, his potato plants waving in the currents, pike lurking in the shadows of his smokehouse.

There is nothing that will so effectively decimate a salmon stock as blocking off a river. This has been well understood for centuries: in England, during the reign of Richard the Lionheart (1189–99), legislation stipulated that any construction spanning a stream must implement a gap the width of a well-fed year-old pig. But such knowledge has rarely been heeded. Despite the decree, the dams of millponds blocked off the streams of medieval England, and when they did, the salmon disappeared. Eight centuries later, Seattle City Lights dam killed 239,000 Chinook fry in a single night when the water level dropped. They are extinct on the Spokane, since Long Lake Dam went up. A poem is carved into the granite there: 'The Place Where Ghosts of Salmon Jump'. The Grand Coulee dam, on the Columbia River, has been called 'the single most destructive human act towards salmon of all time'.

There have been attempts to mitigate these disasters. Some dams have installed fish ladders that allow the salmon to scale their heights by navigating a series of pools. On the Snake River in Washington, fish are herded into trucks and driven to the top of the dam, and then, in the autumn, they are driven back down again. At Rock Island Dam on the Columbia River they are taken in helicopters. A company called WHOOSHH Innovations has a 'Salmon Cannon' that will suck fish from the bottom of a dam and fire them out the top at 20 mph, apparently without side effects, although it is hard to watch footage of it in action without finding it in some way troubling. Nothing is as effective as removing a dam entirely. The Elwha Dam on the Elwha River in Washington, built in 1910, was knocked down in 2012, and the Glines Canyon Dam, ten miles upriver from it, was removed in 2014. By 2016, there were 4,000 king fry and 32,000 coho fry heading from the Elwha out to sea. Pressure is now mounting to undam the Klamath, a river running the California-Oregon border, which has seen abysmal Chinook runs in recent years. It would be the biggest dam removal in American history.

The Yukon's only dam is the one in Whitehorse, one little hiccough near the headwaters, making the Yukon the longest stretch of free-flowing river in North America. Built in 1958, a fish ladder was added in 1960, after a heap of salmon appeared at the foot of the dam each summer. The ladder is longer than the Eiffel Tower laid flat, which only makes me wonder at this final indignity, as the exhausted kings at the end of their run are forced to perform a near endless series of chin-ups, raising themselves eddy by eddy up fifty feet to the spawning grounds beyond. The migrating smolts, when they depart for the ocean, are blasted through the dam's four turbines. A rollercoaster for fish they call it in the visitor centre, albeit one with a 30 per cent fatality rate. Gulls and diving ducks wait downstream to pick off the dazed young fish as they emerge. Others are chewed up by the blades. Others are killed by absorbing excess nitrogen, like a diver getting the bends.

In 1983 a fish hatchery was added in an attempt to supplement supply. It is located in a complex of one-storey buildings behind a chain-link fence over on the edge of Whitehorse, and before I began paddling, at the end of May, I had gone to visit it. I found

Phyllis in there, tagging fry. There were thousands in the tanks, each a shade more than a gram, turning one way and then another as if with a single collective mind, like a flock of tiny underwater starlings. She held each to a machine that fired a millimetre of metal into the cartilage of their nose. Beneath a microscope the implant would reveal where they were hatched, and when, and where released.

'It doesn't look like you need to be that precise,' I said, as she rattled through them one by one and dropped them into a bucket by her side.

'You do,' she said.

This was her twenty-fifth season. She had done nine thousand today, and it was a little after lunch. When they hit the water they floated for a second before shivering into motion and commencing to swim circuits.

'I like knowing that my hands have been on the ones that come back,' she said.

Along with the implant, every hatchery fish has its adipose fin clipped. A small, fatty protrusion of skin between the dorsal and the tail fin, salmon seem to manage fine without it, although it is believed to confer some small benefit to fish when swimming in turbulent waters. But its absence does provide an instant indication when one is caught that this is a hatchery fish. Phyllis let me have a go. It was not easy. The fry lay gasping in my hand as though its batteries were running down whilst I jabbed around with the scissors. I left it with a ragged flap that it would bear for the rest of its life and dropped it back into the water.

Next door is Warren Kapaniuk. Warren is the sort of man not born to office life. Dressed in rubber boots and plaid shirt and baseball cap he seems too large for the room, stabbing away at a keyboard with fishy fingers. Pushing aside a stack of graphs he drops a few salmon eggs into a plastic cup from the water fountain, and pours in a slug of white vinegar.

'These were just fertilized yesterday,' he says.

He hands me a magnifying glass. I peer at them. The vinegar has made the yolk translucent, and there, at its core, is life. The cells have split. There are eight of them, deep within the egg, clinging to each other like a cluster of bubbles. By their next division they will be numerous enough that it will be impossible to tell them apart without a microscope. Warren asks me if I would like to make some salmon.

In the main room, in pools, to the hum of cooling systems, the adult kings turn slow and langorous shapes like ornamental carp. Netting covers the tanks so that they can't spring loose. These fish have been arrested on their trajectory to their spawning grounds, plucked from the top of the fish ladder, and they are hardwired for escape. I follow Warren down the steel walkway between the tanks. He is dressed head to toe in waterproofs, and he carries a white plastic net over one shoulder like a gunslinger. They have selected thirty males and sixty females from the run, a random sample. It would be easy to meddle with the genetics if they had the inclination, selecting for the biggest and the strongest, but it is important to resist. From one of the tanks Warren scoops up a female, and she writhes about, pounding the surface with her one silver fist. He gets a slipknot around her tail and draws it tight, and he holds her up at arm's length, swinging from the rope. She falls as still as a bird hung upside down.

'Have a feel of her belly,' says Warren. 'You can tell when she's ripe.'

I grope her at the midriff. She is soft and forgiving. That feeling, he tells me, means that the eggs have loosened, and that she is ready now to spawn. In the wild all it would take now is for a male to stimulate her instincts. Unfortunately, we are all the males she's got. He takes a wooden cosh that hangs behind him on the wall, holds her so that her cheek lies against his rubber boot, and clocks her sharply on the head. She shivers and the life goes out of her. Her journey finishes here. But her purpose will be realized, the next generation assured. And maybe this is a better end than the decay she could have otherwise expected, the putrefaction of her still-living flesh, or being eaten alive by a bear. He suspends her from a beam by the cord around her tail and prises her gills apart to reveal their scarlet, coralline fans, probing the filaments inside with a knife until the blood begins to run. It runs through sluices in the concrete floor. There is blood on his wedding ring. Blood he says, would contaminate the eggs. We wait until it slows, each droplet forming at the tip of her nose before falling, and then he runs another cord behind her gill plates and through her open mouth, as though bridling a horse. Then he stuffs her gills and her mouth with paper towels, to soak up any drops. She looks dressed for some fetish event, trussed by a shibari artist. He carries her over to where he has placed a white tub on the floor, and he passes her to me. I take her, and hold her in the manner of an accordion, keeping her taut, and Warren rubs along one flank and the eggs begin to flow.

They squirt into the tub in a single, steady stream, an orange so vibrant that it appears artificial. When the flow stops he slides his knife into her belly, slits her from the head down to the anus, and the rest of the eggs come forward in a gush, a mound of spawn so large that I would not believe it could all fit in one fish unless I'd seen it. There are five thousand eggs, he reckons, give or take, about average for a female. Large females can carry eight thousand. The really big ones, the mythic beasts of the past, sixteen to twenty.

The males are in tanks outdoors. Many have already had their milt taken from them once and are very much on the turn. Lawrence Vano is working out here. Lawrence is small to Warren's size, bright-eyed and ruddy-faced. Together they form some improbable duet, a couple of concerned parents working side by side over the salmon. Lawrence shows me the dead fish, piled in a wheelbarrow like the victims of a plague, their skin like marbled paper, brindled burgundy and white and pink the same shade as cheap mince. There are a few gonads slung amongst them.

'I've been here for thirty years,' says Lawrence. 'And I am really done with killing things.'

The heads will be sent away to have the metal strips removed to check their data, the carcasses fed to the eagles. Warren fishes a live male from a tank. The cock fish is far more placid than the female. This will be his last shot at the next generation, and perhaps he already knows it, and is resigned and fatalistic from his previous violation. These men coaxing his milt with their cheap tricks. Warren holds him in a towel and Lawrence gets out the Ziplocs. With a practised thumb and forefinger Warren strokes the male's flanks, from the nose end to the tail. The fish's mouth hangs open, and it squirts a couple of teaspoons of milt into the Ziploc that Lawrence proffers. It is hard to see this as innocuous. It is like those objects in the great museums, objects of Australian Aboriginal art or treasures looted from the pyramids, objects never meant for the gaze of so many, other eyes, all magic stripped away. Warren fills a second bag, and zips it shut.

'We have an individual relationship with every fish,' Lawrence tells me as we walk back inside the building. This is a tiny hatchery; those used to stock the Columbia might raise twelve or thirteen million fish a year. Here they are stocking just four creeks, the traditional spawning sites of the Chinook before the dam went up.

'We used to raise 350,000 fish,' he says. 'Now we do 150,000, and we still get the same returns. There is a holding capacity to every stream. That's what humans haven't really grasped. We just keep on expanding, chewing stuff up. These salmon can't do that.'

The eggs are divided into two Ziplocs, so as not to put them all in one basket. Lawrence passes one to me.

'They've travelled 2,600 kilometres to be here,' he says. 'Don't drop 'em.'

He hands me one of the bags of semen, and I pour it into my bag of eggs. I zip it shut, and give it a shake. Almost instantly there is a reaction. A thin foam begins forming around each egg, microscopic bubbles gently fizzing. I hold the bag up to the light, watching it happen. I hold the beginnings of life in my hands.

We empty the bags into buckets of cold water and the eggs sink to the bottom. The water hardens off their membranes until they become tough enough to bounce. After some minutes Warren swills the water off, as though panning for gold, which it turns out is his hobby, and we carry them through into a side room and pour the whole lot into trays. Here the eggs will pass the winter between stacks of other trays, with water running through them to replicate life in the redds. Warren and Lawrence carry pagers in case anything should go wrong; if the water were to stop, they will have just hours to save an entire year's stock. In a week or so the eggs that have not taken are removed, to prevent contamination. Come spring they will be salmon. We stand next to each other, looking down at our work.

'You're gonna be the dad of five thousand salmon,' says Lawrence. 'Better start shopping. Better get another job. Writing won't support you now.'

Hatcheries have been around since the 1850s, and at their inception they seemed such a wonderful, guilt-free solution to decimated salmon stocks. For every fish taken out, by a dam or a fisherman, you could simply put in one more. Hell, two more, a thousand more, why not? In the second half of the nineteenth century, hatcheries proliferated through Scotland and France,

Canada, New England and the Pacific Northwest. Eggs from the west coast were shipped across the world to restock barren wild rivers, even to Australia and New Zealand. It appeared a panacea. John M. Crawford, Washington State's superintendent of hatcheries, believed hatcheries meant that 'there is absolutely no real reason for the eventual depletion of [salmon] by overfishing or the advance of civilization'. If only ecosystems evolved over several million years were so simple.

In 1896, Livingstone Stone, who headed the project to restock England's rivers, admitted defeat:

> I doubt if there was one person who had heard about it . . . who did not believe that salmon were going to become abundant again in the Atlantic rivers . . . The eggs hatched out beautifully. The young fry, when deposited in the fresh-water streams seemed to thrive especially well. They grew rapidly and when the proper time came were observed to go down in vast numbers to the sea. What afterwards became of them will remain forever an unfathomable mystery. Except in very rare isolated instances, these millions of young salmon were never seen again. What became of them? Where did they go? Are any of them still alive anywhere in the boundless ocean? Or are they all dead? And if they are dead, what killed them?

Whether runs can be restored seems to depend on how degraded the environment has become, and whether fishing can be curtailed. But where hatchery stocks have shored up fish populations, it is because they have replaced wild stocks, not enhanced them. Hatchery fish are used to fighting for their food, and when released into streams they will outcompete the wild fish; they may even eat wild fry. Hatcheries introduce diseases. An over-reliance on hatcheries has allowed fishery managers to turn a blind eye to overfishing and habitat loss. And perhaps most damaging of all, hatchery fish introduce an alien set of genes into a wild population that has evolved to be genetically specific to its natal streams for hundreds of generations. And then there is the expense. Chris Stark, a fish biologist, estimates that each Chinook that returns to Whitehorse has cost Yukon Energy, which pays for it, in the region of $300. Extrapolated globally, this is not a sustainable figure to prop up the environment. We are not much closer to finding answers than we were a hundred years ago. In 1979 the Thames Salmon Rehabilitation Scheme began trying to restock the empty Thames, but in 2011, when the project was terminated, it had nothing to show for its efforts but a bill of several million pounds. Some things that are broken, perhaps, cannot be fixed.

The Tanana, the Yukon's second biggest tributary, drains a watershed the size of Ireland. Where the two rivers meet the silt load becomes yet heavier, the water saws at the canoe. The day is hot, too hot for itself. It has been like this for close to a week now, and a wind is coming, and out over Fairbanks, way off east, the clouds are piling up. The skies are so huge that you can read the coming weather many hours before it arrives. It is the tail end of July, and people are already talking about the end of summer coming.

We are more and more careful, as time goes on, to camp out on the islands. The salmon are in full spate, and as those that are spawned out begin to wash up in the tributaries, bears are making their way down to the Yukon from the hills where they have spent the early summer. We still see them only rarely, but they are forever in our minds. One sat in the shallows, the water up to its chin, cooling off as we paddled past.

It is now the tail end of the king run and only the stragglers remain, the 'big fuckers' Andy spoke of. But now the summer chum are here, flowing east beneath our boat in their spawning colours of green and red, and next will come the pinks, and then the silvers, and then the sockeye, and then last the autumn chum. Most of the kings that have not been caught, or turned off the Yukon up the tributaries to breed, will be in Canada by now. Kings travel fifty miles a day; for us, going downriver, fifty miles is exceptional, and we're still eating three meals a day.

Although that could be what slows us down. Half the boat is full of food. We bake bread and cook tagines, bulked up with wilted dandelions. In the mornings there are pancakes with wild raspberries, in the evenings there are fish chowders and elaborate stews of donated moose meat and of cabbage, cooked over the fire, rounded off with rhubarb crumbles. And then there is the salmon, which everyone wants to share, despite their smoke-houses being half empty this summer. When we stay with people they press it on us, and when we leave they fill our bags: with whole fish and fillets, heads and bellies, smoked and half smoked, canned and dried. One afternoon a man pulls up along-side us in his skiff, hands us two Ziplocs stuffed with dried strips, to welcome us, he says, and races off down the river. We snack on it as we paddle, until we are so oily it becomes part of our odour, and in the evenings we roast it, grill it, fry it, or slice it thin for sushi. I do occasionally consider the ethics of investigat-ing a fish's decline whilst stuffing my face with it. It is in these moments, by the fire in the evening, the day over with, the dishes done, when I feel the journey most acutely: the simplicity of it, of days that feel full, and fully used by the day's end.

Two days out from Tanana we make camp far out in the mid-dle of the river, on an island of sand that rises only slightly above the level of the water. We hadn't seen it until we were almost on top of it, obscured by the gentle swell. No bear in its right mind, we figure, would be making its way out here. There is nothing on the island but for a single dead tree. Ulli sets the tent and I make a fire and we are finishing up dinner when the wind starts to pick up. We see it first in the movements of the swallows, buf-feted through the air. Down the valley, the way we have come, the sky is so grey it is blue. The trees on the banks are stirring, shivering in their canopies, soughing in their branches, and the sounds come to us undiluted across the water's vast expanse. It is hard to make the guy ropes fast on ground as soft as this. We hurry about, piling whatever we can find down on the pegs: dead limbs, the float barrel, the canoe. I wonder if the water is rising. It looks as though it might be rising.

The sky turns, if anything, darker. We watch it, sipping tea. Our shadows stand out stark against the sand. Ancient spruce flex turquoise in the light, bending forward, deferential, snapping back and flapping like those inflatable men outside of garages. The light is as sharp as a knife, carving out each individual colour. The wind keys up a pitch, whipping the sand about, so that it flows about our feet like the ghost of some other river. The thin alder that climb the bluffs flare white. The atmosphere thrills me; I feel electric, animal. Lightning cracks. And then we hear the rain. The sound of it swells as it sweeps down the valley as though a herd of horses were approaching, charging at us across the water. We hurry inside the tent and get the canvas zipped down, and then it hits.

The tent bends and quivers, straining at its tethers. We lie inside the sleeping bags staring up at the thrumming canvas. I've not used this tent in a storm before, and I have a sudden jerk of realization quite how far we are from anything. What an illusion it suddenly seems, this membrane between home and the vast and storming world. Outside the flaps the sky is as black as it has been in weeks. The puddles on the beach reflect it. Thunder barrels across the sky. I try to read, but the storm is too absorbing, there is nothing to do but be in it. We make poor, nervous jokes to each other. A lone goose, barking, is hurled out of the clouds.

It rains all night, and when we wake it is still raining. The wind has died, at least. I push further down inside the sleeping bag, and think I might just sleep the day away. A little later Ulli stirs and takes a look outside. The island is half the size it was. The canoe is partly floating, and the river is metres from the tent. Ulli shakes me awake. The rain hammers on the canvas. We pull our waterproofs on and push out into the weather.

It rains, without cease, for a week. The gear is wet and the tent is wet and we are increasingly wet in spirit. The far bank vanishes in mist. Over this next two-hundred-miles, there are no villages at all. No fish camps, no shelter, nothing. We sit in the tent playing endless games of crib and reading each other's books. Some days it rains so much that we bail as we paddle. Drift is flowing constantly, great trunks of tree at orders of magnitude bigger than any trees round here. Creeks are so jammed it would be easier to run than paddle up them. The boys who glean logs and sell them for firewood, lassoing them from their skiffs as though corralling cattle, will be making a good living, but fishermen dread this type of weather: the drift plays havoc with the set nets, the logs can jam the wheels. The land is sodden, and we are finding it increasingly hard to pitch a tent.

We are exhausted when we find somewhere to stop. It is an island, but a big one, with the willows reaching almost to the shoreline; it is better to be on open ground to give big mammals time to see us before they stumble on the camp. But we have been paddling since breakfast and we are tired and tetchy with each other, and the next flat spot could be hours away. There is a brief break in the rain. Ulli builds a fire with some dry wood we have carried with us, pieces of birch bark for tinder sealed inside a Ziploc. I walk out from camp, looking for Labrador Tea. A small plant with elliptic, slightly succulent leaves, and a burnished, rusty underside, like certain caterpillars, or nails left out in the rain. It gives hot water a soothing, antiseptic taste. It is never far away, which is good, because I get uneasy straying far from camp. I find a few plants, and I take a few leaves from each, and I walk on, looking for more. I am developing some very active superstitions out here, and one of them is that if I pay due attention to the things within my power – if I leave no trace, if I take only a little of what I need from each plant so as not to impede its growth, if I dig a decent hole when I shit – then I am less likely to be eaten by a bear.

There is no reason to this, of course, but then, there is no reason to bears. We have with us an airhorn and two cans of Yukon Magnum bear-spray – industrial-sized cans of pepper spray – and I hold a belief that mostly things work out, but that feels precarious little when coming from a life based squarely upon certainties. I have my senses, and I use them more consciously than normal: I pace the ground, scouting for bear sign when we stop, I sniff around; once, when eating lunch, we both get a feeling at the exact same moment that we should not be where we are, and, feeling foolish, we hastily pack up and paddle off. Who can say? Perhaps the wind changed, or the pressure dropped, or a branch fell in the woods. It seems overly romantic to believe that my sloppy instincts, much dulled from underuse, should straightaway start firing when immersed in the correct conditions. But it didn't take long after the wolves returned to Yellowstone for the deer to remember old behaviours. Percy Henry, in Dawson City, had called the wolf the doctor of all animals. Maybe, for us, it is the bear.

To be seeking new forms of understanding and protection, shorn of my rational ways of predicting the world, should come as no surprise. The settlers who first arrived here found what to them appeared a pre-Enlightenment world, the native knowledge unscientific. No one would talk directly about a bear; it was always 'that big animal'. Certainly no one would speak of a bear in the presence of a woman. Women and children were forbidden to look at them; it could bring terrible luck to a village. In one village I met a newly arrived man from the Lower 48 who spoke of bears incessantly, such was his fear of them: 'One day a bear is going to come knocking at his door,' an elder told me, 'saying here I am if you want me so bad.'

The Karuk of California made their fish spears only from the trees of the tallest mountains, for otherwise the salmon would see them. If salmon eyes are kept in a house overnight then the entire run will disappear. Same-sex twins are said to have a special affinity with the fish. The missionaries wrote off such animist ways of approaching the world as primitive and pre-rational, but the settlers were equally capable of thinking of the salmon as having an interior life that extended beyond the biological. Each culture seems to find in the salmon its own approach to life's Sisyphean struggle: for west coast tribes, it embodies the selfless sacrifice for future generations; for Europeans, it is about a rugged individualism. To the gentleman angler, the salmon has forever reminded him of his own stiff-upper-lip approach, a Charge of the Light Brigade mentality. Izaak Walton, author of *The Compleat Angler*, spoke of the salmon carrying out its 'natural duty', Dickens of how 'he will rush at a cataract like a thoroughbred steeple-chase horse, returning to the charge over and over again, like a true British fish as he is'. It is noble, doomed, regal, wild. An animal with an intimate connection to its homeland, and with the resilience and determination for great journeys; an animal that goes to embrace its death with the open arms of a Zen master. More recently, a symbol for the West Coast activist, a crusader against clear cuts and dams, an individual that triumphs against all odds, conquering the currents of progress.

Each Pacific tribe has rites depicting how the first king of the season should be eaten. The Kwakiutl of Vancouver Island serve its roasted eyes to their chief; the shamans of the Tsimshian people, dressed as fishermen, parade the first salmon through the village on a platter of fresh cedar bark. The Ainu in Japan used a ceremonial club of willow to kill their salmon, and the first of the season was passed into the fisherman's house via a special window reserved for the purpose; rice and malt were wrapped in a bamboo leaf and placed beside the fish's head. In Siberia, the Nyvkh people placed two sticks of willow in the river, at the time of the spring and autumn runs, and launched between them a small boat made of birch bark, filled with offerings of food that included salmon. Each year, in the middle of May, Alaska Airlines fly the first catch of Copper River king from the port town of Cordova to Seattle. The plane is painted like a salmon, tail up, fins out. There is a live blog of their progress. On the runway in Seattle, a red carpet is rolled up to the cockpit, and the pilots carry the salmon down it. Out on the tarmac, assembled chefs compete before the gathered press to cook the winning dish for an assembled panel of fishermen and retired quarterbacks.

I gather enough leaves of Labrador Tea and return to camp and set a pot on the fire to boil. We are walking back to the canoe to unload, and I am looking down at the trail of a moose in the sand, when Ulli says 'There's a bear', and I think what an inappropriate joke, and I look up, and there is one.

I had so long imagined the moment that it feels like a kept promise. It is a grizzly, or a brown, the first that we have seen. *Ursus arctos horribilis.* The day hones in upon it. It is on its hind legs, as tall as the willows, perhaps six feet high, or eight, and with furrowed features it peers myopically towards us. It is maybe twenty feet away. Both brown and black bears can be black or brown: you can distinguish the grizzly by its hump, and unlike the black's muzzle, which is rounded, the grizzly's lengthens to a snout, more collie than labrador. I am not sure if it is an adult or a cub: on the south coast of Alaska, where they grow huge on salmon, the adults can reach upwards of twelve feet, but here, with a diet of mostly berries and roots, they may not make half that. It is so very *there*. The medium of the wilderness serves to level our experiences: we are both on journeys through it, and we have both surprised each other. My body understands the gravity of the situation in a deeply instinctive way.

I had thought, on seeing a bear, that I might be torn, conflicted between fear and the privilege of the moment. It goes right to the heart of my ancestors' experience, but for modern

Western man it is an unusual situation. The promise of violence is familiar, but there is none of the malignancy that would usually accompany it. Because of that, not everyone reacts predictably. Tourists have been seen feeding bears from out of their car windows; in 2012, a photographer in Alaska's Denali National Park was killed after getting too close to a grizzly, on his camera a series of intimate photographs taken from just fifty yards away. But I discover I have not strayed too far from my instincts. As one, Ulli and I raise our hands and make some unplanned, primal noise. It is unconscious, and indeed we have been told never to react like this, that with the grizzlies we should appear deferential. On reflection I suppose that we are puffing ourselves up. The bear is startled. It drops to all fours, and then it turns, cocks its head back at us, and canters off into the scrub, disappearing through the willows with a delicacy that his several hundred pounds belies.

My heart is thumping in my chest. My hands begin to shake. The beach is glaringly empty. We stand there, looking after it, and if it wasn't for its prints left behind in the mud I could think we had made it up. It must have lasted all of ten seconds. A goose honks, and then another. We reek of fear. We pack up in minutes, a task that normally takes hours, our flight mechanism still in overdrive, and we push off, looking for another island. There might be a bear on that one, too, but at least we won't know about it.

We wake late, exhausted, to faint music. It sounds like The Tokens. It is faint enough, or we are far enough from anywhere, that I assume that I must be imagining it. I check the map. There is nowhere on the trip that is further from a village: a hundred miles from Tanana, a hundred miles to Ruby. I look outside the tent. No bears. The weather is still limp and insipid. I must be so starved of music I am hearing rhythms in the river. It certainly sounds like The Tokens. *In the jungle, the mighty jungle.* We lie around camp for most of the day, weary, unable to get going, and occasionally there are more snatches, of The Beatles perhaps, or The Bangles. I am sitting, drinking coffee, when a Cessna rises above the spruce of the island straight in front of us, close enough that I can see the pilot. He waves at me. I lean back, gawping, as he passes overhead.

We paddle out, beyond the island, towards the north bank. By now it's Bill Withers, 'Ain't No Sunshine', and the rain is coming down again. As we round the bend we see, up on the bluff above us, a wind sock pointing south, and three crosses, big enough for crucifixions, silhouetted against the low cloud. We moor up beside a row of skiffs and climb the bank towards the music.

Above the rise the land opens out, and there are children everywhere. They swarm over the grass, chasing footballs and frisbees and each other. There are perhaps a hundred of them, more people than I have seen in one place for a long time. They shriek and holler, and they move with complete abandon, lost in play, oblivious to the weather. The music is coming from speakers suspended in the trees. Log cabins cover the land behind them, some small and others huge. There is a basketball court, surrounded by bleachers, in the centre of it all. I spot an adult, standing over by the trees. Perhaps he's their slave, I think. I wave to him. He waves back. Then he catches hold of the frisbee and takes off running into the spruce, pursued by several children.

We climb the steps of the largest cabin, beside a sign that reads 'Porcupine Hall', and a man comes out to meet us, surrounded by a crowd of children. He radiates health, as though he is emerging from a magazine.

'Welcome to Bible Camp, you guys!' he says.

His name is Joey Katches, and he tells us he is from the

Foothill Christian Fellowship, Meadow Vista, California. A blond child with a bloody nose clutches at his sweater hem, the rest of them are native.

'You guys smell like camp,' says one. Joey rubs her head in gentle admonishment, but he cannot deny its truth.

Joey invites us to stay for dinner. Long tables line the hall, and along them are perhaps two hundred children, arranged roughly in order of size. We sit at a table with some of the oldest, sixteen to eighteen years old, who are helping out with the younger kids. They are from Buckland, up near the west coast, two hundred miles and three flights away. Dinner is rice and chicken and cabbage, and after, strawberry cake, improbably pink. It is Joey's birthday, and we all sing. He looks absolutely thrilled. Over dinner we play a game. After you take a drink of juice, you have to rap your glass twice on the table; if you forget, you do a dare. I am careful to rap my glass. Billy, sitting across from me, a big lad with a deep voice, a face full of scars and a mouth of twisted teeth, mutters that he isn't playing.

'Have you had bad experiences in the past?' I ask.

Billy, not realizing I am making a joke about the game, thinking instead I am on an opening gambit to a conversation that will recall the traumas of his childhood, looks across at me like murder. This is supposed to be a safe place, far from it.

The Kokrine Hills Bible Camp was established in 1965 by missionaries from New York. Since 2009 it has been managed by two Alaskan natives, Roger and Carole Huntington. Huntington is another of those names that ripples down the river, which I will hear everywhere from Fort Yukon to Kaltag. Roger, an Athabascan, was born in Koyukuk in 1944, and Carole, an Inupiaq, was born the same year in Nome. They live year round at the camp, in a two-storey log cabin on the far side of the clearing.

In 1988, Roger had two experiences that changed his life.

That was the year his uncle died, and his aunt had asked him to sing 'Amazing Grace' at the funeral. Roger believed in God; it was a product of his upbringing. He was raised in an orphanage by Catholic priests after his mother contracted TB. Those priests, he says, sexually abused him during his years there. Later, he would lose fourteen family members to suicide. He believed in God but could feel nothing but anger towards him. But his aunt had asked him to sing at the funeral, and he thought he might have one more go at trying to understand just what was so amazing about this grace.

Sitting in a library, day after day, he read the biography of John Newton, the Anglican clergyman, abolitionist, and author of the lyrics to 'Amazing Grace'. He was swept away by it. Roger had found success as a business leader and Indigenous activist, but his life felt increasingly hollow.

'I had a lot of ego, a lot of pride,' he says. 'At one point I had three private airplanes of my own, just for my own enjoyment.'

Later in that same year, he narrowly survived a crash in one of those planes, escaping with third-degree burns covering 60 per cent of his body. He is terribly scarred to this day. Roger and Carole started going to church around this time, and later they took over the management of the Bible camp. The process of their healing has been a long and winding one, and they wanted to offer some of their own experience to children who might have gone through similar experiences. Roger quotes Job 14:7: 'For there is hope for a tree, when it is cut down, that it will sprout again, and its shoots will not fail.'

Whilst Roger and Carole manage Kokrine Hills, almost everyone involved with the day-to-day running of the camp is white, most of them Californian. Delbert Mitchell is an exception. Born in the Inupiat village of Selawik, he first came to the camp in 2010. The camp, he says, is where his heart is, and the only summer he has missed was last year, when everything was cancelled with the wildfires. Roger hopes, in time, that the camp will be managed by natives, and Del would be the obvious choice. Del says he is praying for that. After dinner, he shows us around. He shows us the art room and the mudslide, the beach where they go swimming and do the baptisms. We see the archery range, with papier-mâché targets of caribou and bears, and the new washroom, funded by the Grahams, with the biggest concentration of flushing toilets on the Yukon. I ask Del what it is about this place that makes the kids want to come.

'I can't put it into words,' he says. 'But from what I see it's a loving, caring environment. It's fun, and you get to learn the Bible. It's a safe place, and you don't always get that in the villages. Not hardly ever. Some of the villages are really tough and they're really hard places to be.'

After his first year in the camp, Joey Katches invited Del to come and live with him and his wife, Amanda, in Los Angeles, and Del attended community college there. Nancy, another teenager who I sat next to at dinner, told me: 'I used to speak to Joey and Amanda on Skype when they lived down in California. They really listened.' They often called the kids in Buckland, and in time, with their young family, they moved up there. Nancy remembers how nervous she was the first time that she met them. She didn't know if they would remember her name, and that was about all she had to hang on to. But they called out to her as soon as she entered the church. She has been coming to camp ever since.

Joey invites us to stay the night. In this weather, a bed indoors is very welcome. We are asked to the evening session. In the chapel, we take seats in the pews. Del finds us Bibles, because we do not have our own. Set to an evangelical tune, the service begins with a photomontage of the day, projected against the back wall. Kids covered in mud, kids on the climbing wall. The cook out in the kitchens, laughing behind his vast ten-gallon pots. Kids grinning, kids hugging, kids defying their expectations of themselves. Another of the Californians, beaming, strumming a guitar, leads us in a hymn. *You're rich in love and you're slow to anger. Your name is great and your heart is kind*. Then he says a prayer.

'We're all worthless,' he says, his eyes closed. 'We've all let God down.'

Have I? I think. Perhaps I have. Amen, then.

Paul stands for the sermon. He tells us that God is like a parachute. 'If you jump out of a plane and you don't have a parachute on,' he says, 'you'll die.' His head is shaved, his eyes are wide. 'I'm from Russia,' he says, 'where the Christians were persecuted. They wanted to kill the last Christian live on TV. But you know what? Now there are more Christians in the country than ever before. We will prevail.'

It is all starting to seem a bit much, for eight-year-olds before bed. Paul speaks of evolution. The scientists are in denial he says, they are following only themselves. We can choose to follow God or else we follow ourselves, which is no different to following Satan. And we fail, all the time, to live up to God's name, to follow him: only Jesus managed that. He introduces us to words: omniscient, omnipresent. He raises one hand ceilingward, and intones 'holy, holy, holy'. He holds the Bible out before him, trembling, and says that if we remember one thing from the sermon, it's that we should all tremble before the word of the Lord.

'Tomorrow,' Paul says, 'we'll talk about depravity.'

The water is higher than anyone remembers it being for July. At the Bible camp they marked places on our map where there were flat beaches good for camping, but often when we find them there is nothing there but river, pushing up against the submerged roots of trees. Four days out of the Kokrine Hills and we are still paddling at midnight, unable to find anywhere to stop. We decide to keep on going to the village of Koyukuk: it seems unlikely we will find anywhere to stop sooner. We have been told that Koyukuk is off the main stem of the Yukon, a mile or so up from where the Koyukuk River joins it, and that, to avoid a long hard paddle against the Koyukuk's flow we could take a short cut through Whontleya Slough, emerging on the Koyukuk just above the town. (Story has it that once all rivers flowed both ways, downstream on one side and upstream on the other, but Raven decided that this made everyone's life too easy, and fixed it so that they only flowed in the direction of the sea.) On the slough the banks come close, and the monotony of alder and willow are broken up by flowers and other shrubs. A moose gazes at us as we pass. A great horned owl drops from a branch ahead and glides before us down the channel. It settles on a branch that hangs out over the river and observes us with its feline face. As we drift beneath it the owl takes off once more, wingbeats slow and steady as a heartbeat, and it flies through the thick, close spruce beside us without a sound, as ephemeral as shadow. Each time it perches and pauses, waiting for us, and then it carries on again. A furious mass of feathers, needling us with its eyes. Later an elder will explain to me that although most people believe that owls bring bad luck, they really bring all kinds. People don't know how to hear them any more, she says, but anyway, they speak in Athabascan, so even if people could hear them they wouldn't understand.

'The last time an owl spoke to someone was 1973,' she says. 'And it was to Walter Nelson and his son. And Walter got mad and scared the owl off, and that was the winter that his son fell through the ice. And you know, they never did find him.'

By the time we emerge from Whontleya Slough, several hours later, the wind is blowing strong. The night is a gloomy blue, about an hour away from dawn, and on the far bank the lights of Koyukuk glow. It would be nice to be over there. But the Koyukuk is wider than I had expected, perhaps half a mile across, and now a half-mile is looking like a very long way indeed. We watch the breakers rolling past from the safety of the slough's mouth, the wind gusting spray from off the tops of them. And as we edge down the collapsing bank, the soil dripping with rain, the willows impenetrable, it becomes apparent that Ulli and I must either make the crossing or spend all night sitting here in the canoe. We eat chocolate and discuss what we should do, but there is not much to discuss. If we stay here we will certainly be soaked and miserable by morning. The crossing remains an unknown quantity. Perhaps the owl was a good sign.

They are the biggest waves yet that we have been in. We both bend to it, heaving on the paddles. Ulli is in the stern and rudders us straight so that the waves break on the bow, and in the bow I pull, plunging the paddle deep, willing us forward. We fly across the water. We climb each peak and drop from its top, climbing up the next. We find our stride, in unison. We are taking on water as the waves crash on us, and I am already soaked. In the stern, Ulli is bailing. As we cross beyond halfway we start to believe that we will make it. And I realize, as much as I am scared, that I am also having fun. We shout encouragement to each other. We start to sing. The water feels somehow warm. The lights of Koyukuk draw closer, and finally we make the shallows and pull up on the beach, laughing, euphoric, and saved. A few teenagers are passing a joint a little further down the shore, and they watch us, what the river has just washed in, as we clamber, still giggling, up onto the beach, and start to erect the tent.

We have arrived, it turns out, in time for one of the summer's baseball tournaments. Six villages who get their teams together on the weekends, driving several hundred miles in their boats to play some ball. Nulato, Ruby, Koyukuk, Huslia, Galena and Kaltag. It is a pretext for people to hang out, for families to get together, and for young people to hook up. The boats are wedged gunwale to gunwale along the beach. The rain does not stop. It pools and drips through the tarps over the bleachers. 'K-O-Y-U, K-U-K' chant crowds from the bleachers. 'Koyukuk Raiders here to play.' The hitters slide through the mud, stretched out, the boys chew gum and swagger. The boys look at the girls, furtively, the girls look at the boys. We could be anywhere in small-town America. The Koyukuk Raiders and the Raiderettes are all dressed up in blue, cowboy-and-Indian-style Indians drawn on the backs of their hoodies, feather headdresses and Roman noses. The women of Koyukuk bring continual rounds of food down to the tables, and we all help ourselves. Roast beaver, goose soup, moose heart soup, baloney sandwiches, iced cake, Hershey's Kisses. There is plenty of salmon. There is coffee in an urn.

We stay in town for several days, watching the games. After dark we walk the streets, the streetlights on and the four-wheelers running back and forth, piled high with people. There is a buzz of drinking, of something happening in town. Bonfires burn around the field. The teenagers are hanging out in the washeteria, where they can charge their phones. Down the road, in the community hall, a dance is finishing. People are sitting along benches that run the four sides of the room, and the band is on a small stage in one corner, local boys all of them. There is a wood stove in another corner, and a ceiling of crumbling plasterboard, and it feels as though this hall has seen many nights like this. *Smoke salmon, not cigarettes* reads a poster on the wall. We sit down next to a man who introduces himself as Eddie. I ask him what sort of dancing people do here.

'You find a partner,' says Eddie, 'and that's dancing.'

The band launch into Old Crow Medicine Show's 'Wagon Wheel', an Alaskan anthem if ever there was one. *Oh, the North country winters keep a gettin' me now / Lost my money playin' poker so I had to up and leave.* One by one couples take the floor, holding each other, holding cans of Miller Lite. Someone kills the strip lights. The harmonica howls. I hear old folks bemoaning the lack of fiddle music these days, but it is mostly old folks dancing. The teenagers sit awkwardly around the room, the boys and girls on separate sides. Everyone dances just how they please. Eddie beams, waving his arms around. The guitarist finds his stride in a drawn out solo. The frontman lights a cigarette. 'This is a federally supported building,' shouts an elder over the solo. 'Put it out.' He is caught in limbo, between rock and roll and the respect of his elders, and he inhales long and hard, uncertain, and grinds it beneath a boot. *So rock me mama like a wagon wheel/ Rock me mama anyway you feel/ Hey mama rock me.* Ulli and I dance, and people smile that we are. There is real glee in the air. When the band takes a break everyone goes outside and stands around in gaggles under the street lamps in a haze of cigarette smoke. We walk back to the tent and lie there, sleeping fitfully, listening to kids on the bank in the early light passing a bottle around, shrieking their teenage dreams out to the water.

At last the sun comes out. I watch some baseball and go for a walk, and when I get back to camp in the evening there are a woman and a few kids swimming, the kids all in lifejackets. On a bench on the bank, watching them, are Lucas and Billy. Lucas is twenty-eight, topless and hairless. He carries a bottle of R&R (Rich & Rare, Canadian whiskey) between his bare back and the waistband of his jeans, holstered like a gangster. His nipples stand erect in the wind that is blowing off the river, his skin all gooseflesh and soft fat like a baby's. His eyes sit a little sunken. Billy's smile is easier. He is the little brother of Ruth, the woman in the water. He sits with his eight-month-old nephew on his lap and there is another, half-empty, bottle of R&R beside him. He confides to me it is his birthday, that he is twenty years old. I have just been to buy ice cream for us as a rare treat, but the store only sold it in gallon tubs and I'm happy to have people to share it with before it melts. Billy passes me the bottle in return.

'Take a shot?' he says.

I hold my hand up, shake my head.

'Warms your belly,' he encourages. 'What's wrong with you? Are you a Christian?'

I give a noncommittal shrug. In truth, I have been warned to be careful around alcohol here, and I'm already feeling uncertain about how this evening might finish. Gift-giving is bonding and is supposed to be reciprocated, but I would rather be sober when things pan out. Billy takes a swig and grins. My loss.

'They tell me it don't taste so good when you get to twenty-one,' he says. 'When you don't have to hide it no more.'

'I've been sipping for weeks,' Lucas says, as Billy passes him the bottle. 'Buzzing for weeks. When I get started I just don't stop, you know? I had work out in the woods cutting firebreaks but then we all got laid off. That was maybe two weeks ago. Company went bust, I guess.'

Billy's in good spirits and wants to talk. He wants to know how much trainers cost in London. He cannot believe that the drinking age in England is eighteen. He wants to know if I met his cousin in Fort Yukon. He tells me of his plans to start a cattle farm out here: he wants to make something of himself, he says, he wants to do it for his village. Billy already knows where the land is and what animals he wants. He needs to get it going, he says, before the shit hits the fan.

'Geez,' he says. 'It's the first time I've talked with out-of-towners.'

Ulli asks him why he thinks that the shit will hit the fan. He takes a long, deep breath.

'People take everything and they don't save nothing,' he says. 'We see the world leaders and what's happening to the world. We know we have to look after our own. We're the only niggers out here. We gotta protect our lands. Protect our rivers. No trash here. We respect this place. Have to keep the white hunters out.' He looks at me, considering his audience. 'We'd let you through. You drifters are okay. You don't have a gun.'

Ruth is out of the water now. She has the baby on her lap and she is taking pills, some drug that I have no knowledge of and no name for. A couple, and a gulp of booze. She passes them to the others. The kids crawl about in the sand. The boy, in a Spider Man T-shirt, kicks sand in the youngest girl's eyes using his superhero powers. The girl starts crying. Ruth seizes hold of her.

'You fucking worthless piece of shit,' she says.

The girl cries harder, and runs off into the bush. Ruth chases her whiskey with iced tea, wincing.

The sun is setting, another storm is on its way. The kids suck on popsicles from out the cooler, and drop their plastic sheaths between the beer cans and the cigarettes. The boys tell bragging stories, of being out on boats whilst drunk, of capsizes and near drownings, of illegal fishing, some ancient salmon called a blue dot. Stick-and-poke tattoos twine Lucas' arms, many already fading, his birth date, the sign of Libra, a grizzly, some girl's name. 'Athabascan' in italic script across his nape.

'There's salmon bootleggers,' he says. 'Everybody knows it.' He stares at me, seeing what I make of that. 'I used to fish illegal,' he continues with his swagger. 'Thirty-nine bucks a pound. One year I made seventeen thousand bucks. I sent it all over the world.'

'He's my cousin, but he's full of shit,' says Ruth, her smile slanting.

Lucas ignores her. He pulls out a bag of salmon strips.

'See that?' he says. 'That's twenty bucks, right there.'

He offers me the bag and I take a piece to chew on, happy to have something to accept.

'All over the world they don't give a shit about overfishing,' says Lucas. 'So why here?'

All the villages down the Yukon have different policies on alcohol, voted on by their tribal councils. Tanana is wet: it has a liquor store, although the opening hours are so conceived as to make prolonged bouts of drinking require a little planning. Koyukuk is supplied by a river-drive-thru liquor store called Last Chance, three miles down on the north bank, the last licensed store on the Yukon. Other villages, like Anvik, down-river, are damp. It is legal to possess but not to sell; there are limits on what can be ordered in. In Anvik the limit is eight bottles of liquor, five crates of beer and four bottles of wine, per person, per month. They use forklifts to bring the booze off of the barge. The majority of villages are dry. Hairspray and Lysol are kept behind the counter, and the bootleggers are the market. A bottle of R&R might be $10 in Anchorage, $30 here, $50 in Anvik, and as much as $200 in the dry villages on the coast. These policies work about as well as prohibition worked last time. The higher the price, the more it needs to count.

In 1984, Harold Napoleon, chief of the village of Hooper Bay, a hundred or so miles south of the mouth of the Yukon, beat his four-year-old son to death. He was so drunk he did not remember doing it. While serving his sentence he wrote an essay called 'Yuuyaraq: The Way of the Human Being'. He described prison as a laboratory from which to study the Alaskan natives who had been sent there by way of alcohol abuse: they comprised the majority of the inmates. Harold Napoleon did not believe that the native pre-disposition to alcoholism was a result of malevolent genetics. This once popular theory was based on colonial assumptions rather than biology, developed by white men who believed that they were better at handling their drink. Drinking stories, like history, are written by the victors. But even though this theory is now discounted, I met many Indigenous people who believed it of themselves. The true cause, says Napoleon, is not physical, but spiritual. And if the cause is spiritual, then the cure must be spiritual, too.

In 1900 an outbreak of influenza, brought by missionaries and miners, began in Nome, and swept throughout Alaska. By the time it had run its course, 60 per cent of Eskimos and Athabascans were dead. In 1918 the Spanish influenza pandemic wiped out another half of Nome's population. The *angalkuq*, the shamans, the village healers, the holders of the culture and the spirit, found themselves impotent before its spread. Everything that had sustained the tribes, for thousands of years in one of the harshest environments on earth, their medicines, their healers, their spirituality and their stories, fell apart in the face of this new force. Illness came from invasion by evil spirits, Napoleon writes, and evil spirits that their *angalkuq* could not placate, evil spirits that came to so many of them, must have been evil indeed. Must have been, perhaps, deserved.

A new generation, orphaned from the old ways, grieving, abandoned, rootless, rose out of the ashes. It was no great leap for the missionaries to convince them that the old ways had been heathen. The natives were in shock. They had just buried their children, their parents, their lovers. There was no longer any map. They settled down, they stopped dancing, and they prayed. When the children spoke the words of their old language their mouths were taped shut by the incomers, or washed out clean with soap. Harold Napoleon writes of how people internalized their guilt, their pain, the unmentionable things that they had seen. He compares their situation to the young men who came back from Vietnam.

Alcohol had arrived with the whalers; much more came with the Gold Rush. For the dark nights of the soul, it was the only medicine that remained, the only semblance of purpose in an otherwise unstrung world. Disease and famine lasted into the 1950s. Alcoholics raised abused, unwanted children who became alcoholics in their turn. And despite improvements in material conditions from the 1960s onwards, the symptoms of the fallout have perpetuated. Alaska has three times the national rate of rape, almost six times the national rate of child sexual assault. Half of Alaska's women have suffered sexual or domestic violence. Alaskan natives have the highest suicide rate of any ethnicity in the entire United States, and higher than any other country in the world.

'Have you been to Fairbanks?' Billy says to me, his eyes muddled now, leaning forward in male conspiracy. 'You seen that Latina puss there? Ho-lee. That half-Latina puss. I wouldn't mind hitting that half-Latina puss.'

His eyes are far off, distant.

'You want some puss?' he says to me. 'I can find you a girl. Last girl I had, she used to come out in my boat with me and suck my dick while I was fishing. Get down on her hands and knees like that. Real nasty, bro. Real nasty.'

Ruth spits out her cigarette smoke. 'You should shut up because you were probably related,' she says.

We laugh; the tension breaks for just a moment. Billy blinks at us, considering.

'You're all laughing at me,' he says. 'If you laugh at me more I might get in a fighting mood.'

I go and sit by the tent where Ulli is, apart from them. Billy keeps looking darkly over at me. Ruth has her hand on his chest, soothing. They are talking but over the wind I cannot hear them. The sound of the baby crying comes from the direction of their truck. Lucas comes over and hangs an arm across my shoulders, pulls me in, proprietorial, his face a shade closer than it should be, an unlit cigarette stuck to his lower lip that beats time as he speaks. He smells sweetly of sweat and of smoke, his hair limp, a sheen to his forehead, and he points to a scar down on his side, by his belly, a wicked puckered thing.

'See that?' he says. 'My friend shot me. My best friend shot me.'

He stares at me. I am lost for words.

'But I survived,' he says. 'Shot him back. Dropped him. Killed him.'

'I'm sorry,' I say, weakly.

'Got off, though,' he says. 'Self-defence.' His eyes flicker. 'It tears you up inside, when you kill a man. It still haunts me every day.'

I can't name the threat, but it is there, as though a thing of actual substance. He lets his arm fall from my shoulders.

'But we don't want to talk about that,' he says.

He steps back, lights his cigarette, watching me intently to see what I might say. Something is expected of me.

'That must have been awful,' I hazard.

He shouts a laugh, and slaps me hard on the back.

'I'm joking,' he says, with those eyes still fixed on me. 'You don't think I'd kill someone. Do you?'

I smile and try to laugh along with him. He meanders off, and when he returns he is in a child's pink towel dressing gown to try and keep out the cold. He looks like a boxer pre-fight. I lose track of what is happening. I don't know where the kids are. Billy staggers up to the tent, the previous slight forgotten. He is intent on finding me some puss. He leers at Ulli.

'You hitting that?' he says.

He asks me if I would like to swap her for his cousin. I am scared and mutter some limp reply. He piles great logs onto the fire, and it spits sparks into the brittle brush and in the direction of the tent, the growing wind fanning them up. When I try and adjust them he snaps at me.

'A fire is like a baby,' he intones. 'You gotta feed it.'

Lucas has dozed off. A skiff drifts past, having slipped its mooring, and a couple of minutes later another skiff comes chasing after it. The men are shouting, laughing, drinking.

And then suddenly it is raining. It pockmarks the sand; the fire hisses like it's wounded.

'Holy cow,' says Billy, gazing into it.

Ruth pulls Lucas awake. Lightning cracks the twilight. The three of them stumble for the shelter of the truck. Ulli and I make for the tent and lie there, willing the engine to start. At last it does, and it ticks over for a while, and then we hear it drive away.

Later on, in the dark, I wake. Headlights illuminate the tent.

'Hello?' moans Billy. 'You there?'

I do not answer.

'Where'd my sister go?' he says.

Through the tent flaps I watch him, lit by a four-wheeler, his drenched T-shirt stuck to his skin, standing there as though he has forgotten even the most basic facts about himself. He stares about. And as I watch him, he pulls a plastic bag from out his pocket, sinks to his knees, and fumbling through a haze of rain and dark and alcohol, he starts to gather up the trash.

It is a week since we left Koyukuk. The Yukon shifts, prowls across the continent, stalking the sea from different vantages as its moods and fancies take it. It worries at its banks, and where the permafrost has warmed them the earth comes away in chunks. The splashes and their echoes drift to us across the water. As the trees approach the banks they become unstable, their roots dangling in mid-air beneath the overhangs, so that seen from the river it is a cross-section of the land, the world as some underground creature might see it. They lean out at increasingly implausible angles, as though dropping from the end of a conveyor belt, protruding so far out, clinging on only by a handful of lateral roots, that we can canoe beneath them, a phalanx of trumpets raised in welcome.

The air smells damp and fecund, the earth's innards exposed. Peat black and dripping, a grubby sheen of ice running through it in a layer. How simple seem its inner workings now. As when one skins an animal, and how quickly it surrenders the illusion of being a living thing, a rabbit or a moose, and becomes nothing more than flesh. We pass the Palisades, where mammoth tusks poke from the earth, and on the wind hangs the stench of decay from Pleistocene mammals exposed to the light after a hundred thousand years. The vastness of this place seems to leave it open to destruction. For banks to be carved off, for bluffs to slip, for swathes of fire to run through it, all scarcely impacting on the landscape.

On the opposite bank, on the inside of the meander where the current slows, the silt that has been drawn from the tops of mountains a thousand miles hence is replacing what is lost. In the old days, Holy Cross sat on the right bank of the Yukon, and paddle steamers dropped anchor in front of the village. Now the river is a half-mile walk away, and the only water access to the village is a long and sluggish paddle down Ghost Creek Slough.

Three sandhill cranes lurk in the long grasses of the sandbars that have formed at the slough's mouth, heads bobbing above the underbrush, the bright red of their crown against the white of their cheeks, Japanese flags with beady eyes. Their calls bubble up through their long necks, warning themselves of us. They are the oldest bird on the planet: their fossil record dates back two and a half million years, twice as far as most other living birds, and to see them is to better understand the link between birds and dinosaurs. They move with the rangy swagger of lizards, and when they take off, no longer able to tolerate our presence, they are so vast and so ponderous that the mechanics of flight appear less of a mystery, as though they are hauling themselves, hand over hand, into the sky. We watch them, the three of them, as they gain height, still warbling and rattling, to go gliding across a mile of water and land on the far bank.

In the slough the banks are close, and we float beneath bowers of cottonwood. We cast out the line and drift. When the current slows and the silt sinks it is possible to fish on a rod; in here, the fish can see the twinkle of the lures. A belted kingfisher sits at the top of a dead spruce. With the water so high the river leaches out into the woods. Bright orange bracket mushrooms grow from the stumps of dead willows; an empty bottle of vodka drifts amongst the trees. Ahead of us there is a disturbance in the water, a thin ripple drawn by something moving just beneath the surface. Some fish, presumably. And I am just wondering what sort of fish cuts through the water like that when it conks into the side of the canoe.

On the rebound I can see it clearly. It is a salmon, a pink, what the locals call a humpy. Or that is what it was. We had first seen pinks a couple of days ago, rounding the cliffs at Bishop's Rock, and they were there in the vast eddy, thousands of them, their humps cresting the water as though the river itself were boiling. But those were healthy, flashing silver, bound upriver. This fish is monstrous. Its body is the colour of eyes gone blind, not white itself, but as though seen through a veil of cataracts. Its eye stares back at us, unblinking, as vacant as the rest of it, and on its flanks are ugly grey sores where the skin hangs loose and ragged. It is grotesquely humpbacked, the portion of its back between its head and its dorsal so round that there is as much fish above it as below. Its kype is caught in a rictus somewhere between a snarl and a leer, and its gills are clouded with fungus, where it gasps for air as though breathing through cotton wool.

The pink is the shortest-lived salmon, on a strict two-year cycle. It is also the smallest, the average weight just under five pounds. I see another, and then two more, and then hundreds. The further down the slough we get, and they are everywhere. Clouds of fish, puffy, airy, scabrous, drifting across the surface. A few of them dying but most of them dead, pale and rigid, just a tail or a nose protruding from the water. Tangled in the submerged reeds, faces locked in disbelief. They have great welts around their eyes and noses, all along their backs. Many are already growing algae, a mossy green that blurs their outlines. Where beavers have dammed the slough the navigable channel narrows to a few feet, and here they have washed up in droves. We push through them with the paddles, and they clunk against the sides as we cut a wake, and the smell is of fish markets and death. At the end of the run, their purpose realized, no energy left to fight disease, they will drift like this in a state of gradual decomposition, their lives not ending in a final moment but a gradual blurring into death.

In Holy Cross we camp beside the slough and the bank is

thick with rotten fish and other trash and the whole town smells of it when the wind blows a certain way. The pinks have not spawned here. They have drifted down from somewhere else, and they have never showed up in Ghost Creek Slough before. No one knows what they're up to, what this means. We wake to a heavy mist. It is 10 August, and an old man outside the post office tells me with authority that the fall is here already, and that the winter will be hard, before shuffling off into the gloom. Four-wheelers pass on the edge of vision. On the border of Athabascan Indian and Yup'ik Eskimo territories, this town is a mix of both people. The steeple of the church, Russian Orthodox in design, presses up against the underbelly of the clouds; Russian missionaries first reached here in the 1840s. In her front room, from where she runs a tour-guide business with her husband, and where two men from Oregon despondently watch the Olympics while they wait for the weather to lift so they can fly off to catch pike, Connie Demientieff tells me that if I really want to meet someone who knows fishing on this river then I should go and find her mother, Mary, eighty-four years old. And if you can get hold of a bottle of Chardonnay to bring her, Connie says, then all the better.

We push out from Ghost Creek Slough. Mary Demientieff's fish camp is half a day's paddle downriver, set back in a wide bay beneath Tabernacle Mountain. Hills ring the bay, but from this point on, high land will be scarce. Since Koyukuk, 250 miles before, the river has run south. Despite the ocean at times being no more than forty miles away, the Yukon has been kept from it by the ridge of the Nulato Hills that run parallel to its western bank. But now, at this camp, the hills come to an end, and the river can bend around to the west and strike out on its final 250 miles across the delta. It is this bend that forms the good eddies for the set nets. We pull in beside two other boats, both skiffs, at the foot of a fallen bank. There are several cabins, some ply, some log, a smokehouse built from scraps of corrugate, all thrown up haphazardly across the patch of land. Washing blows on a line. Standing above us, watching us, is a little girl, deadly serious, in the way that some little girls are. She has seen us coming, and meets us with instructions on where to moor. I climb the bank and tie the bowline to a flagpole, where a flag embroidered with the words *Live, Laugh and Love* snaps about like a flame in the wind. With all the gravity of diplomatic ritual, the girl hands each of us a two-inch piece of celery with peanut butter daubed into the hollow. These are small canoes, she explains, for eating. We are represented by the peanut butter. My thoughts turn again to the bears.

Mary Demientieff is walking across her yard towards us. She is dressed in slippers and a red smock, a blue flannel turban to cover her hair. She walks with her hands behind her back, her whole body angled forward, and she moves across the yard as though toppling forward, a continual upsetting and restoration of balance with each step. Her knees are not what they used to be. Her eyes have the same glint as her great-granddaughter's, her cheeks round and bright, her face softened but not wrinkled. On one wrist she wears a band with the words *Padre Pio – Pray Hope and Don't Worry*. She beams at us and opens her arms.

'Adventurers!' she says.

Mary spends her summers out here at camp, her winters in Holy Cross, as she has done for over sixty years. A creek falls down from the mountain top and meets the river, splitting the shore in half; on one side is Mary's camp, and on the other is Jeffrey's, one of her sons. Mary has six sons: Jeffrey, Joseph, Jareth, John, Julian, and Bergen. She has six daughters, too: Carol, Connie, Cathy, Correine, Candyce, and Lisa. By now she has lost count of her great-grandchildren, but they number somewhere around a hundred, and then there are her great-great-grandchildren. Her family, her life's work. Sometimes people will look at her askance when she tells them the number of children that she had.

'I tell them "I was doing my duty, as a wife,"' she says. 'They think maybe I was oversexed or something.'

Bergen is out here for a few days, up from Anchorage, to help his mother out. The little girl is Mary Junior, his granddaughter. Bergen has the distracted air of someone busy with his own internal commentary. He draws out the syllables of my name as far as they will go, and fires thin brown streaks of spit, like punctuation, as he talks, the by-product of the tobacco jammed inside his upper lip. We sit around a table inside one of the cabins. Mary makes tea and puts fish roe and fried bannock down in front of us. She has family the length of the river, the breadth of the state, and she wants to know where we've been, who we've seen, how the fishing goes, who's at camp. She is delighted that we stayed with friends of hers, the Honeas, up near Ruby. Once again, I am astonished by the ready hospitality, the ease with which people open their doors and offer beds and food. We are offered so much food that Ulli has long since given up not eating meat, and she has fallen to the moose and beaver and bacon, with pancakes and birch syrup, with a vegetarian's zeal. Besides, this is the food of her childhood.

The walls are lined with flattened cardboard boxes to keep in the heat of the wood stove. Beside it there are shelves of dried goods, and the brands that have become familiar by now: Dunkin' Donuts coffee, Smucker's strawberry jam, Jif peanut butter, Crisco, Sailor Boy Pilot Bread Crackers. Outside, Mary Junior swings back and forth, crosswise across the hammock, singing a song of her own devising. Everywhere there are icons of the Virgin Mary: an ornament on the kitchen table, framed paintings on the walls, a plastic statue on the front stoop, her hands spread wide, as though demonstrating the size of a fairly disappointing salmon. There is a shrine up on the hillside, topped with the steeple of the original church of Holy Cross. On the wall above the stove is a black and white portrait of the founder of the Sisters of Saint Anne, her face framed by her apostolnik, a crucifix around her neck. *Mère Marie-Anne Fondatrice des sœurs de Sainte Anne*, says the inscription below. *Née le 18 avril 1809 décédée le 2 janvier 1890.* I ask Mary who she is, and she tells me that if I want to know, then I must know some other things first.

Mary Demientieff was born Mary Dahlquist, in Nulato, in 1932. Her father, Bill Dahlquist, was from Stockholm, but what had brought him to Alaska Mary didn't know. She never knew too much. He died when she was five. She thinks, maybe, he had something to do with sailing. She remembers pictures in the house, old photographs of great wooden boats beneath a canopy of sails, prows dipping to the waves, and she remembers, too, him coming home from his work in the jailhouse in Nulato, and sitting in the kitchen in impenetrable silence. He was her mother's third husband, third of four. The first died in an avalanche. The second mauled by a bear. No one divorced in those days; they didn't last long enough. She remembers her mother. She remembers her crying, always crying. And she remembers her drinking. She remembers the stories that she told her.

When she was seven she was taken to the Holy Cross Mission Orphanage, the same orphanage that Roger Huntington of the Bible Camp would be taken to, and abused at, ten years later. Mary went on the recommendation of one of the priests in Nulato. Her father was dead, and her mother was incapable of looking after her. She was out of hand, the Jesuit priest said, and her mother was not in the habit of questioning the opinions of the church. Her mother stood on the riverbank, her eyes raw and red, as she watched her daughter shipped off to where she herself was raised. Mary went downriver on the paddle steamer

S.S. *Nenana*. For most kids the expensive trip on the steamer was one way, and they would not return home until they were eighteen. The boat sits in a park in Fairbanks now, a museum to itself, like much of the rest of Mary's life. One of her grandsons plays piano in its saloon for the tourists.

She remembers all that like it was yesterday. Turning up in Holy Cross in the middle of the night, September, and the air sharp with the fall. Smoke rising from the houses, the leaves on the ground. A farmer in a tractor collected them in a wagon, and they drove down a track lined with spruce and in the distance she could see the lights of the mission, lit up like a pinball machine. She thought she'd died and gone to heaven. There were three hundred children in there. Her and all the other orphans. Mary felt lucky that she still had a mother. She even came to visit her, once.

They were raised by the bell. The bell told them when to get up, when to lay down. When to eat, when to wash, when to work, when to stop talking, when to start talking. Ten years, she was there in the mission. Ten years to make her into the woman she is today. They taught them the value of discipline.

'I'd be on Skid Row now,' says Mary, 'if it wasn't for those nuns.'

But it was hard that the nuns never held them. Her mother might have been an alcoholic, but at least she hugged her. But then, the sisters had no children of their own, so how could they know? They used to skate with them in wintertime, in their flowing habits, gliding across the frozen river like some kind of exotic waterfowl. Some of them weren't much older than the kids. There are photographs from the time of groups of native children, some in suits, some in parkas, and in the back row, out of all context, looms a nun in her habit and wimple, like a phantom figure that was not seen when the photograph was taken, and appears only in the development.

They learnt the four Rs: reading, writing, arithmetic, religion. English was insisted on, any other language was that of the Devil. The nuns took them to Paimiut, a village a few miles downriver from this camp, and it was there that they learnt to cut fish and the boys learnt to set nets. Every month they would be given different chores. Scrub the pots. Do the laundry. Weed the gardens. Sweep the stairs, and for the love, there were so many stairs. Funds were meagre, and so the pupils maintained their own buildings and grew all of their own food. Mary liked working in the infirmary, with Sister Mary Edward, and she liked being the sacristan, helping with the altar linens. Most of all she liked looking after the little ones, the new orphans, too young to learn still, but brought to the mission with no place else to go. She would dress them, and she would hold them. Just hold them, and breathe in the smell of their scalps, and for a moment she could be somewhere else, somewhere softer, where it was love and not obedience that could make you into someone better. Annie was alive still, up in Anvik. Annie's first memory was being held by Mary. And that's something, Mary thinks. That's something.

One night, when she was fifteen, she ran away. There were three of them, Mary, Ursula and Irene. They snuck from their dormitory in the middle of the night and followed the hill down towards the lights. The mission might have given the village of Holy Cross its name, but that was the extent of the connection. The children were forbidden to mix with those who lived there. Freedom was a thrilling idea. Yet they stood on the banks of the Yukon, dreaming of what it would be like to let it bear them off, and there was no place else to go. They knocked on a door, and Mrs Simms chivvied them inside and gave them akutaq, Indian ice cream, made of seal fat and white fish and berries, and if she wondered what three stray children were doing out in the middle of the night then she didn't say. Mary had never seen anything like akutaq; she thought she was going to be sick. But they'd been brought up by the nuns to eat everything on their plate, and to be honest they needed little encouragement, because they were always, always hungry. The big girls – that's what they were called when they got to a certain age, big girls – would help with the cutting of the kings, and the big boys caught the fish, but they scarcely ever got to eat it. Mary didn't know what happened to it. Maybe it went to the mission's benefactors. The orphans got dog fish, and once in a blue moon, on a Sunday, a little piece of king, a couple of salted bellies on a saucer, laid side by side like two rashers of bacon from a piglet. Such tokens, such memories of a time before the mission, only served to make the hunger worse. It was in the convent that Mary learnt to steal. She sewed a pocket into her petticoat, and in there she would hide the bits of dried fish that she was able to filch when she was working in the kitchens. Once she found a vat of peanut butter, donated by the army after the war, and she snuck into the pantry between chores and ate it by the handful. She got so sick she couldn't touch it again until she was in her seventies.

From Mrs Simms the three of them went next door, where Old John Simms and his wife lived with their children, children who were the same age as them but who in every other way were as different as could be. They stared at each other across the room. The girls realized that they knew no one, and that freedom, after it lost its shine, came along with its own set of problems. They slept beneath a bridge and awoke chilled in the dawn, and they decided to go back. The kids were standing in file, ready for school, and they slipped into the ranks, thinking how clever they had been, at least, to get away with it. It was a secret to keep between them till the grave.

'Ursula Ellanna. Mary Dahlquist. Irene Solomon. Mother Superior will see you now.'

The summons came in the afternoon, in the middle of

arithmetic. The Mother Superior was behind her desk, her face long and thin within her habit, like one exposed segment of an orange.

'Where were you last night?' she said.

There was silence in the room.

'Down there,' one of them mumbled.

'Down where?'

'Down there.'

It was Sister Joanne who shaved their heads. Mary watched her reflection as she ran the clippers across her scalp. It made her think of dogsleds cutting through the snow. And it brought back to her the first memory she had, wrapped in blankets in the back of the sled, their three dogs pulling and her mamma running alongside. It's rare for anyone to remember back that far, but she did. Those early April mornings with the night's crust on the snow, and all the families heading out to spring camp. She remembered waking in the tent in the morning by herself, and the sun coming in, and outside she could hear her mamma's voice talking with the rest of them, all in Indian, and her mamma coming in when she heard she was awake. Her mamma called her by her native name, but she couldn't remember that now. Her mamma was skinning muskrat, the skins stretched out on boards and the tails in a pot. Her mamma said that she couldn't eat the tails, because when she grew up and tried to thread a needle she would shake, because the muskrat tail shakes. But she took the fat and the gristle from the skin, slivers of it sliced with her knife, and popped them in Mary's mouth. She could still smell the spruce boughs that lined the floor of their tent, smell the warm canvas, still smell the muskrat, and even now, nearing the end of her life, when she had to think of something good, in the hospital when they gave her a shot, say, that's the time that she would think of, and it made everything not hurt.

Their hair fell about their feet in drifts. Their heads were bald and white, ugly and misshapen. She thought she would die of shame. The other children kept their distance as though she were contagious. Ursula and Irene had curls and it grew out soon enough, but Mary's hair was long and straight and she bore the shame for months. She wore a handkerchief over her head for so long that it became hard to imagine her without one. But she deserved it, of course, she knew that. She was out of hand.

The pot of tea is empty. Bergen wants to go and check the net, before it gets too late, and he asks if I want to go with him. Ulli stays at camp, and the rest of us head out in the boat. On the far edge of the bay, Tabernacle Mountain drops down directly to the water. The current at the base of the cliffs runs swift as the water urges over the deep rock bottom bed.

'This is the way to travel,' Bergen cries through a stream of tobacco juice. Mary Junior is wearing a lifejacket decorated with characters from *Frozen*, and she stands behind Bergen, who has the wheel, while Mary rests her hands in her lap in a chair at the back by the helm and gazes out across the water. When she smiles her eyes glisten, and sometimes she seems very close to tears, as though the line between joy and sadness has eroded over the course of her life, and has now become very fine indeed.

The net is marked by an orange buoy. Bergen hands the wheel to Mary and she takes it, sets her lips, and holds the skiff against the current. Her confidence isn't what it used to be, but she can still do it if she has to. She drove the boat down to camp this spring. She doesn't like to bother anyone. Bergen pulls a pair of gloves from the pocket of his Helly Hansen dungarees and stands on the bow, hauling the net in hand over hand. I help him. It is heavy work. The net is weighted with rocks, tied around the bottom of the mesh, and it is weighted, too, by the humpies that are so numerous that it seems more fish than net.

'They're so *provoking*,' says Mary, raising her hands to the sky. 'Yesterday we pulled out 159. 159! For the love.'

She finds the photograph on her iPhone, of a tote brimming with them and spilling out onto the deck. Most of these pinks have been snagged by the gills, some by the teeth, and they are bound in Gordian knots that Bergen can negotiate in seconds and that I only make worse with every tug and yank. The ones still alive stare up at me, their mouths agape, speechless at this final injustice. I bind them tighter. I feel like a deranged puppeteer. The nylon cuts into their gills. Their flanks are striped where they have fought against the net, and their bodies are sheathed with a viscous slime that makes it harder for predators to get hold of them. I try to get hold of them. They are soft and spent, and the mucous coats my hands, so that when I part my fingers they appear amphibious. I get one eventually, firmly round the belly. It feels like a cheap pillow. And I hold it just so, massage it in just the right places, that it ejaculates all down my leg.

This is hysterical. Bergen struggles to breathe. I'm not sure how often this happens but it feels like a particularly amateur mistake. It is something I wish that other people hadn't seen.

'Need to get yourself a pair of Hellys,' Bergen gasps.

I yank it free and toss it in the tote, where it lands with a wet clap and shivers and is still. The mound of fish is building. Some, for moments, recognize the gravity of their situation, and beat wildly, ricocheting around the cooler, leaving bloody imprints where they impact. The force makes me nervous, as though the impetus of its wildness is only apparent when held within constraints, like a bird trapped in a room. But as quickly as the salmon start they stop, and lie there panting, an eye fixed skyward, their flanks rising and falling. Mary Junior rests her hands on the side and peers into the tote, entranced by these slabs of muscle that are half the size of her.

When the net is at last empty Bergen feeds it back into the

water. There are several dozen humpies in the tote, and just three of the silvers that Mary wants. Away from the net we hover in midstream, and sling the humpies, two at a time, back into the river. Mary Junior hoists them in both arms, the front of her anorak slick with gore. The silvers are left at the bottom of the tote. We motor back to camp.

Mary got married at seventeen. He was ten years older than her. She told the nuns she wanted to marry and she wanted someone who would treat her well. It was the quickest way out of the convent. They chose Joe Demientieff, from Holy Cross. He had gained a good reputation for taking care of his ageing mother when his siblings had all left. Mary and Joe sat face to face across a table in the dining room, and the Mother Superior sat there at the end of the table, pretending to read Saint Augustine. I want children, Mary said. And I don't want no drinking. Joe said he thought he could manage that.

They were married in the fall, and they stayed married for forty-three years. They left the mission and they went in Joe's boat and he took her up to his fall camp, and the night came down and it started to snow, and there she sat, far from anyone she'd ever known, far from the hundreds of girls, far from the nuns, further from her mother still. Alone, and only this unknown man for company. Joe scarcely ever spoke. But Mary thanked God for giving her a good man, and she didn't think that you could ask for more than that. They built a cabin together that first winter, whipsawing the lumber that drifted down the Yukon, and there was sod for the roof and she caulked the chinks between the logs with moss. Joe was her teacher now, and he taught her everything: how to trap, how to skin, how to can, how to butcher, and how to be a wife.

They had a baby every year. Most of them were born in May, as much a part of the yearly cycle as spring break-up or the king's return. Many were born out here at camp. The older kids helped to raise the younger ones, and all of them had their chores to do. Joe sold his furs and made fish nets, and somehow, Lord knows, they made it. Nine of their kids they sent down to the high school in St Mary's. They had no money, and they paid their fees with moose meat and the potatoes that they grew.

Only one child died, and for that, Mary says, she can be grateful.

'To replace the one I lost, I adopted this one.'

She points at Bergen, who is carrying totes down to his skiff, making ready to leave. He is opening the shop in Holy Cross in the morning.

'He's got relatives up in Kaltag. That's where I scooped that little shit up. There was a priest then, he used to fly his airplane. Father McNeil. He was kind enough to bring me up there. I was having children every year, and it was fine. I loved every one of them. But this one. I fell *in* love with this one. I don't know what kind of love it was. Maybe pity.'

'Why couldn't his parents look after him?' Ulli asks.

'Lot of drinking, up around Kaltag,' Mary says. 'I gathered all of my kids and I said, you and me, we will raise this baby. Because I took this baby without my husband's permission, and all hell broke loose on the Demientieff side. They still haven't gotten over it yet, and he's fifty years old. My husband asked me, you better bring that kid back, that was the way he put it. You'd better bring that kid back. Gently, like. I said no, I am not. I was bullheaded, and I was not going to let him go. All his siblings are dead now. Drinking. Drowning. I don't boast, but they know. Upriver people all know it. I don't mean to brag. God was good to me. And I'm disappointed that Bergen's on so much dope or pot or whatever the hell they call it, but that's the way the world is today. He's still a hard-working boy. He raised that kid. He built his home. But that darn dope. It's like how they don't know how to drink. They have to drink the whole bottle.

I like a glass of wine. But these young people, they don't know how to do things.'

She stares out at the river. 'Look at it,' she says. 'Going and going.'

The expressions flicker across her face like firelight.

'I said no,' she repeats, almost to herself. 'I wasn't going to return it. I was mother, father and grandmother, until he was seven years old. Then they accepted him. So you see why my heart is so heavy. Oh mercy, it was tough. Lots of crying. But quietly. I was a strong person. And look at me now! So old!'

She bubbles up into laughter, a squeal of it, like springing a leak.

'I've been very blessed,' she says. 'And I hope you guys will be like that too. Don't be selfish. Have children. You'll be happy for them. And before you know it they're grown up and they're gone.'

That evening, with the sky fading, we sit in the cabin that Mary uses for her bedroom, a small fire in the burner. The evenings are getting colder. The outdoor walls are painted a jolly, Caribbean turquoise. The three salmon are out on the slab, covered with a wet cloth. She will deal with them in the morning. She doesn't have the energy like she once did. She spent eight years out in Bethel, near the hospital, while they tried to figure out what was wrong with her lungs, but they never found anything and in the end they let her go home. This is the first year that she's not made the climb up to the shrine.

'Listen to me!' she giggles. 'Old women's problems! Old Frances, that's Joe's mum, she never complained. She lived to be 101, and she worked hard until she dropped. When she had a stroke and couldn't use her hands much she tied the fish bundles with her teeth. Never complained. Geez, I should be ashamed.'

Ulli has her guitar in the canoe and she goes outside to fetch it. Mary learnt to play music in the mission. The piano and the guitar, the mandolin and the harmonica, all by ear. Nothing native, of course. No one really knows the native songs any more. The kids are only interested in guitars and country music, Hank Williams, all that stuff. Stuff for drinking to. Mary's fingers are clawed from a lifetime of needles and nets, and the strain of the bar chords comes through in her voice. And although she says she will remember nothing, when she starts to play she plays with glee, unselfconsciously, and the music and the words barrel each other along:

There's a salmon-coloured girl
Who's set my heart awhirl
She lives along the Yukon far away.
Her skin I'd love to touch
But I couldn't touch it much
Because her fur-lined parka's in the way.

She stops playing, laughing and wheezing. She gets out of breath quickly. She had some people come and stay the other night and they really got a kick out of that song. Geez, they were up playing and dancing all night.

'I'll tell you kids a story,' she says. 'My mum told me this story, about a man and a woman that went out to spring camp, just like we did. This man and woman were shooting muskrat, and when they had got enough the man says "I think I'll take the skins down to the village and sell them." And she said "Well you could." And so he loads up his canoe with all the furs and takes off. And when he's gone, this robin comes down to her.'

Mary makes the sound of the robin, shrill and piping, with Athabascan words. 'And the woman says "What the heck, what's this bird on about?"'

Mary sings the robin's song again.

'"Your man's not gone down to the village to sell skins,"' the robin says. '"He's fooling around, fooling around, fooling around." And so she gets her canoe and goes down round the bend and there he is making big party, so she gets his canoe and put it on the fire – it was birch bark canoes in those days. I tell the kids the moral of that story is to be faithful.'

She bursts out laughing, flapping away at the laughter with both hands.

The guitar lies across her knees. She is looking out of the window. It is about as dusky as I have seen it since the beginning of the trip, the far bank almost lost to shadow.

'Joe had that, what you call it?' she says. She shakes her hands in imitation, her Padre Pio wristband trembling.

'Parkinson's?' I say.

'That's it. Parkinson's. I took care of him, long time. He had to go away at the end. He died over there in Anchorage, but I was with him. I asked him a question on his deathbed. Because he never did talk anyway.'

'What did you ask?' I say.

'I asked him just one question. And it was: "Why you never let me go up to bury my mother?" Because it bothered me. He didn't let me go up to bury my Mum. Instead he chartered a plane with his mother and they went to Bethel, down there. Well he went down for a good reason, he was selling his fur, when I think of it now. But back then I was very selfish and I wanted to go up. And they waited for me with the funeral, and they waited for me, but I never came, and they buried her without me. And I held it against him for a *long* time. So I ask him, "Why you never let me bury Mum?" He looked up at me and he said "I don't know."'

There is silence in the cabin, just the crackle of the fire.

'Do you miss him?' Ulli asks.

She thinks for a moment. 'No,' she says. 'I don't miss him. He was a good man. But I had to be very, very strong. Strong in my faith. I always turned to the missionaries. I would talk with them. Let it out, let it out, let it out. So far I made it. But it's still very fresh, still all there. He's in heaven now. Better off than I am. I wonder how it'll be come judgement day. We'll see each others in heaven, maybe.'

'Did you love each other?' I ask, and I don't know why I dare to ask it, but that I feel emboldened by her, and she answers as though it is the most normal question in the world.

'I didn't,' she says. 'I don't know if he did. But it was never brought up. We never said to each other "I don't love you." We just kept doing our thing. *Every* day. Work, get up, work, all day long. We were polite to each other, the children saw that. They're good. I didn't make a show of, oh, I don't like him. No. I was brought up to love everybody.'

'Did you meet anyone after Joe?' Ulli says.

'Yes, I did,' she says. 'He was a schoolteacher. And I thought I fell in love. But then I found out after I stayed with him for a little while, he was a different kind. He was the kind that like other men. So, he asked to separate. I had a hard time. I think that's why I got very sick. Bless his heart. He was a good man. We lived together for a little while. We travelled a lot together. But I didn't know that. That's not right.'

She holds us to her, enveloping us. 'Goodnight, babies,' she says.

There are few photos of the priest George S. Endal. But there is one of him when he first arrived in Alaska, in 1936, standing in front of the mission in Holy Cross where Mary was a girl. It is a sunny day, and glare hits the camera from the windows behind him. He is dressed in his cassock and clerical collar, and his hair is sharply parted to one side. He is smiling awkwardly, his lips twisted by his squint into the sun. It is tempting to try and read into those eyes, to wonder what his thoughts were as he posed for this photograph at the beginning of his sixty-year career in the service of the Catholic Church. Soon after his death, 26 Alaskan natives would accuse him of sexual abuse, and a further 112 would accuse Joseph Lundowski, his assistant from 1949.

Endal moved around the state. From Holy Cross he went to Mountain Village, from Mountain Village to Alakanuk, from Alakanuk to Dillingham. There he met Lundowski, and in the early sixties they moved together to Nulato, before Hooper Bay, and then finally to St Michael, up the coast from the Yukon's mouth. He told his victims that if they told anyone then they would go straight to hell. In St Michael it is estimated that 80 per cent of the children, almost an entire generation of Yup'iks, were abused by Endal and those who worked alongside him.

Upriver, in Nulato, we had gone inside the old mission, still there on the north bank of the river. Books leant together like hands in prayer: *Roman Martyrology*, *Practice and Perfection of Christian Values, Volumes 1–3*, *Enchiridion Symbolorium et Definitionium*. Sheet music, written in pen and ink: 'Holy Night', 'Coelitum Joseph', 'Salve Regina', 'Alma Redemptions'. A stack of dusty vinyl, and beside them, on the player, the needle scoring through the dust, *Trini Lopez Plays and Sings There Was a Sinner Man*.

'Everyone knows the boys in town that were abused,' says Joyce, who showed us round, and who had been a pupil there. 'But no one talks about the girls.'

There was a weight to the atmosphere that was almost tangible. Wooden desks with boards for kneeling, lined up against one yellowed wall. A statue of the Virgin Mary held her child close. In another room, another Jesus, pushed up against the back wall and hemmed in by a bed on its end, peering through the springs. An organ. Brass candlesticks. Linoleum floors.

'Unhappy time,' Joyce said. 'A really unhappy time.'

She still had a Bostonian accent that she acquired from the nuns, like a facet of her soul. Stairs descended into the black hole of the basement, collapsing and boarded off. She peered down there, remembering.

By the 1860s, the 'Indian Problem' was being spoken of by government in both the United States and Canada, the problem being that there were still Indians. Genocide had only got the settlers so far, and for countries now weary of war a new solution was necessary. Nicolas Davin, a politician in western Canada, summed up the new strategy in 1879: 'Children should be removed from their homes, as the influence of the wigwam was stronger than that of the [day] school, and be kept constantly within the circle of civilized conditions where they would receive the care of a mother and an education that would fit them for a life in modernizing Canada.' Typically, these schools were funded by government and managed by the Church. Children learnt to read and write in English. They learnt American and European history, from a Western perspective, and they learnt Christianity's principles. Under the headline 'Young Prince of Moosehide Off To School', the *Dawson Daily News* in August 1911 writes of how Chief Issac was 'elated over the prospects of his youngster being educated', although Chief Issac himself does not sound so sure. 'My boy go school eight year,' he is quoted as saying. 'Me no see for long time. Maybe my boy forget my language. I dunno.'

Madeleine Jackson, an elder who I met in Teslin at the beginning of this journey, is one of two left alive in the village from the first batch of children who were taken. Trucks showed up and loaded them in and shipped them off like cattle.

'When they took us there they took our clothes away and gave us school clothes,' she said. 'And once you start talking your language they strap you with a leather belt that wide and that long. The more you cry the harder they hit you. And you gotta bite your tongue not to cry no more so they wouldn't give you a spanking. Ellie Ambrose was only that big, and she know our language only. And I try and tell her talk English now Ellie, talk English. Just to hear her, a small kid like that. And I can't do anything, because if I try to help her I get strapped. About a month, maybe two months after that she finally realized. She come to me and she say I know what you're telling me. I'm going to try hard to talk white man now. Now she can talk white man but she only talk a little bit Tlingit.'

There are letters from the time, shaky handwriting in broken English, from desperate mothers to distant men:

'*I am the mother of Maggie Linklater and I would like to have her home for summer Holiday. She has not been home since she went to school, and we would like to have her home once.*'

'*I do not want her to stay at school till she is 18; that is too long; when they are too long at school they won't have anything to do with us; they grow away from us.*'

'*I am broken up you sending my daughter's little babies to Carcross. It's too much for me how this is what you want to do.*'

As a local Canadian paper explained to readers in 1917: 'It is frequently quite a difficult matter to induce the parents to part with their children, especially for a number of years, but this natural prejudice will, doubtless, gradually be overcome.'

Unlike the United States, Canada has since gone some way in trying to repair the damage done. A Truth and Reconciliation Commission was established in 2008 (just twelve years after the last residential school closed in Saskatchewan) and concluded its work in 2015. Based on the South African model for redressing the wounds of apartheid, the Commission travelled across the country, collecting statements from approximately six thousand First Nations people who had attended the boarding schools. Of the 150,000 children taken over the course of the 120-year period that they were open, at least 3,200 died in the schools, and of the 70,000 pupils who are still alive, over 35,000 cases of sexual assault have now been brought. Stephen Harper, then Prime Minister, issued a formal apology in 2008, and in 2015 the Commission concluded that the government had 'separated children from their parents . . . not to educate them, but primarily to break their link to their culture and identity . . . These measures were part of a coherent policy to eliminate Aboriginal people as distinct peoples and to assimilate them into the Canadian mainstream against their will.'

Madeleine Jackson pointed out that no one has ever apologized to the parents. 'It was tough for them,' she said. 'Only way they would show their emotion is when they get drunk, and they start crying, and they say why did they have to take my kids away? It's not until they took the kids away from home, that's when everybody start drinking. Everybody, even the women. All over. The whole Canada, I guess, was like that. And all the women and men start drinking and drinking and drinking, because it's the only way they can forget their pain.'

This is the ongoing legacy of the project to 'get the savage out of the Indian'. Alcoholism, suicide, shame, community breakdown, cycles of abuse.

Duane Aucoin, Madeleine's nephew, is a generation removed from those that were taken to the schools.

'I didn't suffer within a residential school per se,' he said. 'I suffered within the residential school that came to my community, that came to my family. So instead of me being dragged away, it was dragged to me.'

It is estimated that over four thousand women and girls have disappeared in Canada since 1980, but there are no accurate figures. After forty years of campaigning for it, the National Inquiry into Missing and Murdered Indigenous Women and Girls is now travelling across Canada, but it has a much narrower remit than many would like, and many people believe it to be reopening old wounds far more than it is healing them.

The trauma of the schools was further exacerbated by a culture in which conflict is traditionally avoided, and showing anger is seen as a sign of weakness. In Yup'ik it is *nallunguarluku*: 'pretending it didn't happen'.

'You don't talk about your hardships,' Debbie Nagano had said to me in Dawson. 'That's how we're raised. When my uncles came back from the residential school they were nothing. And you wonder why a nothing person becomes an alcoholic. How come the kids don't listen now? How come they're into drugs? Well that's why. They see their mother and father drink. Why in the hell not? Who cares about fishing? Who cares about hunting? Who in the hell cares?'

Duane Aucoin told me that for the number of generations it took to reach this place of hurt, it will take the same number to get back. He speaks of the Tlingit phrase Haa Kusteeyí.

'It means our way of life,' he says. 'How we live, how we think, how we do, how we carry ourselves. All that was strong before residential schools. Our elders are helping to bring back Haa Kusteeyí, but there's a lot of our younger people who are still learning. They're having to try to purge themselves of the Western ideas of 'just me', the individualistic viewpoint. They're starting to realize there is a bigger community, and that community extends beyond the current generation.'

Such gatherings as fish camp, he says, are crucial for this work.

Mary wakes early. She lies in bed and recites 'O Divine Heart of Jesus', as she has done every morning since she first went to the mission, three-quarters of a century ago. Its rhythms and the sibilance of its whispers are as familiar as the constant of the river:

O Divine Heart of Jesus,
grant I pray Thee,
eternal rest to the Souls in Purgatory,

the final grace to those who are about to die this day,
true repentance to sinners,
the light of faith to pagans,
and Thy blessing to me
and to all who are dear to me.

She checks her iPhone to see if any of her family have birthdays today, and then opens the door to her cabin and steps out onto the grass. The porcupine sees her and takes fright, waddling off towards the bushes with its hips swinging. She has no love for that porcupine, not since it started crapping in her outhouse and scratching on the walls of her cabin at night, no one for miles around and her mind racing with fear. It comes here for the dandelions, and then it chews up all the gumboots. Just like a spiny monkey. And they just *look* at you, with their quills coming up, ready to shoot. She's seen dogs looking so pitiful, with a face full of quills, her holding them down while Joe pulled them out with pliers. No, she had no love for that porcupine. It's what they call survival food. Nasty little thing.

She walks across the yard and goes into the kitchen. She is humming one of the old mission songs, *Holy Cross, my only cross.* She sets some eggs on to fry and slices some spam into the skillet. For thirty-three years she cooked in the school canteen in Holy Cross. She was there so long they named the kitchen after her. She saw two generations of kids grow up, and her greatest pleasure was to see the kids come up of the parents she had served. Gee, she'd think, I've seen this one before. They all look the same at that age, the sons just like their daddies.

Her sister calls, from Wasilla. They talk about the weather, like they often do of a morning. Her sister says that the leaves are falling already. The middle of August, and the leaves already falling. She says that the earth has turned on its axis, and that everything is early this year.

'It's going to be an early winter,' Mary says. 'That's alright, then. Nothing I can do.'

They were starting to get long-drawn-out falls, not like before. These days it's all messed up.

'Maybe it's the end,' says her sister.

Mary chuckles. 'I don't know,' she says. 'Nobody knows those things. Nobody but God.'

Everything changes, Mary thinks as she hangs up. There used to be a hundred yards of riverbank out the front of the camp, and now it was all caved in, the Yukon coming closer every year. One way or another something would come for her, when it was her time. But maybe not just yet. What was it Bergen said about her? The last of a dying breed. She catches sight of herself in the mirror. She looks more weary every day she sees herself. 'What are you still doing here, Mary?' she says back at herself. She's got no kids to feed now. Her fingers hurt so much she can hardly cut the fish. But still she can't stop coming here. She was given a good life, thank God.

The coffee perks, and she pours a cup. The kids still like to come back home, when they have the time and when Mary pays the airfare. Those times at camp were the happiest of all. When it was just her and her family, and Joe out setting net, and the kids free to run where they wanted. That's when they laughed. It was hard work, but they laughed. And now there weren't even any fish, except for those darn humpies. She hadn't even bothered fishing for kings this year, the regulations were so complicated. Do this, do that. By the time you got the net in you had to get it out again. It's so sad. There's other fish of course, and she should be grateful for that. But there's nothing like a king. Everything changes.

She takes her coffee and her breakfast, and she goes out to sit on her chair by the bank. Maybe the water has stopped rising now. Her binoculars hang around her neck. She wonders how many summers she has left until she'll no longer have the strength to make it back here. She eats her breakfast, chewing slowly, and she looks out at the river.

A day's paddle down from Mary's camp, Paimiut is where the nuns brought the children from the mission, to teach them the skills their superiors had deemed necessary for a Christian life lived out in the bush. It stands empty now. From the river the houses are hidden entirely, the willows encroaching, and the only sign of a human touch is a small cemetery on a hill above the village with the Virgin presiding over it, and a tattered Stars and Stripes fluttering at the top of a pole. There is a depression in the grass where some big animal spent the previous night. Amongst the willows are a few scattered oil drums, affectionately known as the Alaskan state flower. Up the rise a few corrugate shacks at different inclines, russet with rust.

I walk up the hill to have a look around. A scale calibrated up to a hundred pounds sits in a shack with no roof that was once used for the cutting of fish. A wooden mask, some sort of totem, its features smoothed out by the years, is nailed to the lintel. A vast birch spreads itself out over a small cabin. It is the biggest tree I have seen in weeks, and the underbrush here is different, too, not yet reclaimed by the bush, but still made up of the plants that indicate human habitation: chickweed, nettle and dandelion. Inside the cabin, threadbare curtains blow in the breeze from shattered windows. There is a fetid, ammoniac stench, as though something has denned here, the foam mattress all chewed up. The clock is stopped at twenty past four, the calendar stopped at August 2011. There is a tapestry of two bears fishing salmon fixed to one wall, a tapestry of elk in the mountains on another. *God Bless This Home*, in a frame. A box of battered videos, *Fantasia*, *The Nutty Professor*, *The Red Balloon*. A family photo, mounted on tinfoil, is stuck above the window, alongside an icon of a saint. Beside this photo there are others, the daughter now in glasses like her father.

We cook up some of the fish that Mary gave us. The sun is beginning to go down, the sky beginning to pink like a Chinook on the turn. Geese are flying home in sixes and sevens. There is a sense of time having peaked and now unspooling, drifting in reverse. What I know is its inexorable forward march, more people, less land, more stuff. Here, the land is on the march, clawing its way back as the people lose their grip. It is tempting to sense such forces as malignant, but really all they are doing is pricking the ego in ways to which I am unaccustomed. I had thought I would find it comforting, but really it is humbling, and slightly terrifying, to understand that the land is more significant than oneself. To know that nature will one day overrun the work of man even more surely than man has overrun the work of nature.

We eat the fish, and we turn in for the night.

The man told his string stories the traditional Yup'ik way, pulling on threads that he wove between his fingers to make mountains, a fox, a bird taking off and forgetting its legs. He told the story of the two hunters, the one who hunted only for his family, the other who hunted for the whole community. He told the story of a hunter in a canoe, paddling upriver, and as he spoke he tugged on threads so that a boat became an oar. And the hunter had been travelling all day and needed to urinate and he didn't want to stop and he wondered what to do, and the man's fingers wove a cup as he spoke, and then finally, shaking the strings from his wrists and tugging on a loop, what was unmistakably a penis. He collapsed into laughter, his crow's feet bunching.

The kids asked him for the story of the black fish. It was the one he always told. He told them that he was too tired, because he wanted them to beg him for it. It was nice for an old man to feel wanted.

'So there was a black fish swimming up the river, looking for a fish trap to swim in to,' he began, when he could no longer keep up the pretence. 'Cycle of life, right? And as he's swimming up the river he's singing this song.'

The man sang, his fingers twisting the string of his previous stories. He sang the fish's song in Yup'ik. It was untranslatable: the explosive Ks and Qs and the guttural rhythm of his native language gave a feeling to the song that the English couldn't capture. The kids knew it, they had heard it many times, and they sang it along with him.

'And the black fish,' the man continued, 'came to a fish trap that was broken, and some of the fish in it were dead. The black fish stuck his head out of the river to see who it was that owned the trap, and he saw that the village was dirty, and that the dogs were not tied up, and the woman came out to throw out the scraps of a fish dinner and he watched the dogs fight over the bones. The fish did not want his bones fought over. And so he carried on swimming upriver.'

The man sang the song again. He told how the fish came to another fish trap, and this one was well looked after, the fish in it still alive. Sticking his head out of the water he saw that the village was kept clean, that the dogs were all on chains, and that a woman was moving between them ladling scraps into their bowls. And so he chose to swim into that trap.

'And he is caught, and eaten, and you have to know that the consciousness and understanding of a fish resides in its bones. The bones are given to the dogs, and they go down the throat and into the stomach and through the intestines, and finally it comes out as dog shit, frozen in the snow. The snow melts in the spring, and somebody steps in the doo-doo.'

The man rocks back, laughing again, the kids laughing along with him.

'I have a lot of stories like that,' he says.

Lenticular clouds suggest wind high up, but down here it is calm and warm, and in the middle of August these days are increasingly rare. I catch a first smell of the sea. It is nostalgic, magnetic, thrilling. As the raven flies, the Bering Sea is still more than a hundred miles away, and the course of the Yukon makes it further still. It seems improbable that I should smell salt, but I am starting to give more credence to my instincts. The Yukon-Kuskokwim delta is one of the largest in the world. The size of England, for a population of 25,000. (The Ganges River Delta is a little bigger, with a population of 300 million.) It bulges from Alaska's western coast, a relatively new growth two and a half thousand years old, from when the Yukon settled on its current course and began discharging the ninety million tonnes of sediment that it transports every year. The delta is Yup'ik territory; people who refer to themselves as Eskimos, not Indians, who share genetic heritage with the Indigenous of Greenland and Siberia. Once the split between the Yup'ik and the Athabascan was as stark as if a wall were built between them: millennia of bloodshed, a no man's land where delegates would meet to trade seal oil for beaver meat, walrus ivory for mink fur. Once Athabascans warned their children that if they did not quiet down and go to sleep then the Yup'ik would come and get them; on the delta, they told the same stories in reverse.

The Yup'ik call the Yukon *Kuigpak*, and refer to themselves as the *Kuikpagmiut*, 'the people of the big river'. It is indeed a big river. It has always been a big river, of course, but since leaving Mary's camp it has taken on an altogether different scale. It is joined each day by a succession of tributaries that are huge rivers in their own right: the Khotol, the Innoko, the Chulinak, the Andreafsky. 'Neither pen nor pencil can give any idea of the dreary grandeur, the vast monotony, or the unlimited expanse we saw before us,' wrote Frederik Whymper, the British artist, not far from here, in 1866. 'The artist will understand me when I state that it would be necessary in a sketch of this river to make its width out of all proportion to its height, and therefore as a picture it could not be satisfactory.'

At times the Yukon is three miles across, and Ulli and I float on a perfect image that seems to encompass the entirety of the sky. It is a long time since we have set foot on the south shore. Ruby, 570 miles from the mouth, is the last village on the south bank. In the winter the northern bank gets the most of the meagre sun, and with all the villages on one side, it obviates the need for dangerous river crossings. It might take an hour for us to get from one bank to the other, and the wind could do anything in that time. On stormy days we hug the bank, but when it is still we paddle far out to catch the fastest current, a current that is slowing as we get nearer to the sea. Out in the middle, I have sudden vertiginous moments. There is no reason to capsize unless we do something profoundly stupid, but without being melodramatic, if we capsize we will die. It is a terribly long, cold swim to either shore. I feel like a tightrope walker. This fear seizes me for moments, and then I relax again, forget our smallness in the hugeness, eighteen feet of boat on a three-mile-wide highway. We found a copy of *The Old Man and the Sea* in the tribal offices in Nulato, and sometimes we lie back and drift and read it to each other. A juvenile glaucous gull breaks from its flock to circle us. I look up at it, caught in the sun. It dips its inside wing and turns, holding us at the centre of its circumference. It keeps with us. It is smiling at us, I swear.

The landscape has changed. The hills are covered by yellow tundra, the horsetail and other scrub dried out and parched. Lightning cracks over far off mountains. The season is turning. We see it not just in the weather – the more muddled clouds, the growing wind, the atmosphere's weight – but in how other things are shifting, things I would have no awareness of if I were not immersed in them. We see moose everyday now: a lone male; a lone female; a female with two calves. They all seem to be down at the river's edge, and the locals are asking us about their movements. It is nearly hunting season. There are colonies of geese astir and I wonder, is it because they are readying to leave? We have seen two big flocks of Canadian geese in the past two days, each clattering up from a sandbar at our approach and wheeling away in wavering lines, great billows of them, before coming to settle again, streaming back to the pool they rose from like water washing down a plughole. The end of our journey is approaching. And it is getting dark again. It was hard to believe it ever would. Unaccustomed to it, it makes me edgy. I lie in the tent, listening to the geese out in the darkness. A constant bass drone, and individuals yammering above the rumble. Their presence is deeply comforting.

We pass Marshall without stopping, drifting past its waterfront, and it feels like a bustling harbour town, the sun cold and bright, and the men lounging in moored skiffs, repairing nets, talking the day away. A mess of buoys and broken nets, of tumbling tin roofs and satellite dishes. Dogs bark, kids wave, running along the beach to keep pace with us, hollering the same two questions over and over.

'Where you guys going? Where you guys from?'

'England!' I shout.

'England?' one shouts back. 'How far upriver is that?'

A mile before the village of Pilot Station is the salmon sonar project that has been run by the Alaska Department of Fish and Game since 1989. We pull into the eddy and moor up. There is nothing in the state, nothing perhaps in the world, that has such a long-term data set on salmon. Yet there is nothing about the camp that looks permanent. Each construction, and there are ten or fifteen different cabins, is little more than a timber frame covered with plastic sheeting, stretched taut, stapled down. Everything but for the frames is temporary; it all gets packed away at the end of each season and shipped to a warehouse.

'We tried leaving the sink here one year,' says Kyle Schumann, 'and the kids shot it full of holes.'

Kyle is project leader at the camp. It is both the most upmarket fish camp and the most homespun government operation that I have ever seen. Kyle is responsible for most of the amenities: there is a washer-dryer, a propane shower, a vegetable patch, at least ten different types of hot sauce. This is his second season at Pilot Station, his seventh on the Yukon. Kyle is the sort of man who never gets cold in shorts. He wears a shell on a cord around his neck, and a cap that has seen many summers, which reads *Simms Fishing Products* over the brim. Kyle loves fishing. If Kyle isn't shooting ducks, he's fishing. We sit by his computer as we talk, and the screensaver scrolls photos of his life: a photo of a labrador by a river in the autumn; a photo of Kyle holding up a huge halibut; a fish tagging experiment; the labrador in the bow of a canoe; the labrador with a duck in its mouth; Kyle beside the labrador, grinning, in his younger, thinner days. Kyle gets one week off in the summer, and with it he goes fishing. When the season ends he heads back to Montana for the duck hunting, and a winter typing up results. Getting paid to play with fish, he says, is about as good as it gets.

On a board are the daily and cumulative totals of all fish that have passed the sonar since the count began at the end of May. The cumulative total stands at 4,425,623. Beside it are some words in Yup'ik that Donald, one of the men from Pilot, is teaching the rest of the crew.

'*Ataq,* mother. *Aduq,* father. *Tingmaaq,* goose. *Guopuk*, sleep long. *Nijuliuk*, trouble-maker. *Milukuyuli*, (expert) rock thrower.'

The camp is here because it is the one place on the lower Yukon where there are neither sloughs nor islands, where the river flows all in one channel, and as such they have the best chance of counting each fish that comes past. It is about half a mile, bank to bank; on hot, slow days the sonar crew like to swim it. But even though this is the one straight stretch of river, the project is far from simple. Once they worked with paper printouts and a pair of scissors, now everything is digitized, and I suppose I had assumed that, in the twenty-first century, with our reams of data and precision instruments, something as apparently simple as estimating run sizes would not present a challenge; but counting fish still remains one of the most problematic aspects of the entire management of the Yukon.

But then, the salmon has foxed scientists for centuries. For a long time it was assumed that their tremendous leaps were achieved by taking their tail in their mouth and firing themselves like an arrow from their own bow. Hector Boece, the Scottish historian, writing in 1527, was one of the first who got a good grasp of the basics: 'Alwayis, at the first streme of watter that risis, thay discend to the see,' he wrote. 'Thay grow in mervellus quantite, and, with maist fervent desire and appetite, returnis to the samin placis quhair thay wer generit.' A century later in his *Compleat Angler*, Izaak Walton wrote of experiments to tie ribbons around the tails of young salmon and finding fish with those same ribbons, in the same streams, several years later on their return. Yet it was not until the early twentieth century that scientists broadly accepted that salmon had both the intelligence and motivation to return to their natal streams. Even then the question remained of how they did it. In 1946 the freshwater ecologist Arthur Hasler was out hiking near his home in Utah when he came to a waterfall and was overcome by fragrances of columbine and moss that evoked a deep nostalgia for his childhood. It made him wonder if salmon might experience the same. His subsequent experiments confirmed it: salmon that had their olfactory organs damaged were unable to home, whilst fish that he blinded had no trouble in making their way back. Like most other fish, salmon smell in stereo, and they are able to navigate by following the specific mix of amino acids and other chemical compounds from the mouth of their home river to their birthplace. Yet despite these advances, there is still much that is misunderstood. Just how they detect those chemicals. Exactly where they spend their adult lives. What affects their great fluctuations in numbers. How they are able to find the mouth of the Yukon from their feeding grounds far out in the Pacific. It is known that when they set out for their rivers from the ocean they travel in the water's upper layers, so it is commonly assumed that they navigate by the stars, as well as by using particles of magnetite in their skulls that allow them to orientate to the earth's magnetic field. 'That an animal is actually able innately to accomplish something for which men require instruments, charts and tables,' wrote Hasler, 'is quite remarkable.'

I ask Kyle where he lays the blame for the decline of the Chinook.

'Environmental factors won't affect the fish too much in fresh water,' he says. 'But in the ocean. They're in the ocean two to four times the amount of time they've spent in the fresh. There's a lot going on out in the ocean.'

There are various shifts in climate that some believe are impacting on the Chinook. Early breakup of rivers resulting in smaller than normal, more vulnerable, juveniles heading out into the ocean; changing ocean currents that may be spreading disease; shifts in other species in the food chain upon which the salmon depend; increase in migration mortality from the stress and energy depletion caused by warmer waters. Late in 2013 a persistent high-pressure system over the Gulf of Alaska kept out the fall storms that typically churn the ocean and disperse its latent heat. A body of warm water began to build, ultimately encompassing offshore waters as far south as Mexico. It was dubbed The Blob. In some places ocean temperatures climbed higher than had ever been recorded. Fish starved. Krill all but vanished. It was one of the biggest bird die-offs in North American history. Sea lion pups hauled themselves up Californian beaches, emaciated. One was found on a chair on a hotel patio, another slumped in a booth at a seafood restaurant. The Blob persisted, and in 2015 it was followed by The Bloom, as algae spread through west coast waters and failed to dissipate. Crab and shellfish fisheries shut down as toxins hit levels thirty times higher than what was considered safe. Anchovies were deemed too dangerous to eat. Porpoises, seals, seabirds got sick. Fin whales and otters washed up dead, all along the Alaskan coast.

Connections are difficult to prove. There is only an alignment of weirdnesses, a constellation of hunches. The Blob may not have brought on The Bloom. The Bloom may not have killed Alaska's otters. The scrawny salmon that began staggering back to their rivers in 2015 may have been underfed for other reasons. Oceans are complex beasts. The Blob was likely not caused by climate change, but by shorter-term climatic factors. Yet Nate Mantua of the National Oceanic and Atmospheric Administration believes climate change could be 'the puppeteer controlling things in the background'. Oceans are warming. Species ranges are changing; so are the currents. Recent research predicts that the extremes brought on by the Pacific's natural cycles will become more frequent as the climate continues to warm.

Whatever the causes for the decline in salmon numbers, those regulating on the Yukon have the advantage of having seen, time and again, how not to manage their resources, both in salmon fisheries across the northern hemisphere and in other fisheries in the state. There are stories of crabbing captains on Alaska's Bering Sea taking off their gold watch at the start of the season and tossing it in the harbour, knowing it would be small change to buy another one at the end. That was before the crabbing collapsed. As such, the king run on the Yukon, hanging in the balance as it does, has a wider import than just this river: maybe this once, this last time, there can be found a better way to do things.

At another desk on the far side of the office Lindsay sits in front of a screen, clicking one blip after another on the readout. Each blip indicates a fish. Over the years there have been improvements in the sonar technology, and each time the technology improves, predictably enough, they find more fish. Fish and Game then revise the historic numbers for the fish they assume they missed the first time round. Once the fish pass Pilot Station they are not counted again for more than a thousand miles, until they pass the sonar at Eagle, on the border. All this means that there are many people, fishermen and biologists, with a certain lack of faith in the numbers.

Despite the technological improvements, the sonar on each bank do not reach the centre of the river. Fish and Game say that very few fish swim up the river's central channel, and that they have used sonar hung from boats to prove it. But elders in the villages believe that it is the fish in the middle channel, in the strongest current, which are the ones that are bound for Canada.

'I've never heard that,' says Lindsay when I tell her.

There is profound distrust both ways. The Western scientific approach has typically found traditional, Indigenous knowledge to be unsubstantiated, anecdotal. Traditional knowledge has found it hard to stomach that twenty-five-year-old Californians with PhDs turn up to tell them how their land works.

A case in point: in 1977 the International Whaling Commission (IWC) determined that the population of the bowhead whale in the Arctic had hit such a critical point that they placed an immediate moratorium on its subsistence harvest by Inuit and Yup'ik Eskimos, without any consultation with Indigenous populations. In the battles to save the whales that raged through the 1970s, these tribes were the low-hanging fruit. Whilst industrial whaling fleets continued to decimate populations elsewhere, the ban in the Arctic served only to take away one of the most important food sources, and a millennia-old cultural practice, from local people. After much pressure from Alaskan whaling communities, the IWC implemented an annual quota of twelve whales, a quota that the Eskimos declared inadequate for their needs, and one that they said was based on a gross underestimation of the actual whale population. The National Marine Fisheries Service, counting from the air during a spring migration, estimated a population of 2,264 whales: the Alaskan elders said that there were many more than that. Further censuses were carried out during the eighties, and by 1996, using wider aerial transects, along with acoustic sampling, the estimate was raised to 8,200 individuals, and the quota upped to an average of sixty-six whales per year. It had taken twenty years and millions of dollars to prove what the elders had said from the start: that the bowhead population was sufficient to allow whaling communities to pursue a traditional subsistence harvest. It is little surprise, with stories such as these, that many villagers will roll their eyes when the scientists turn up with their iPads and their instruments. I hear about a Yup'ik man down on the coast who harpooned a beluga and found three chum salmon in its stomach. This was unusual enough that he mentioned it to a biologist. But beluga don't eat chum, said the biologist. Well then, said the man, they must've had a real hard time swimming up his asshole.

Todd Brinkman, who is investigating how climatic changes are affecting subsistence harvests in the Fort Yukon area, believes that these two separate paradigms have the potential to complement each other.

'The local knowledge is very holistic, more focused on the interactions of the parts, whilst science is separating the part from the whole and studying the heck out of it,' he says. 'I see a lot of the traditional local knowledge as being much more astute and aware of the changes that are occurring, and asking the right questions the right way. And then I see Western knowledge mostly with their tools of being able to address these changes, and understanding what might be driving them.'

Todd believes that the two could better meet each other halfway by Fish and Game providing some flexibility in the regulations when it comes to subsistence harvests. As unseasonable events become the norm, strict regulations affixed to quotas and dates can seem unfit for purpose. Moose season used to coincide with the first of the cold weather. Now, with warmer autumns, moose are on the move less, and are therefore harder to find. There are more leaves on the trees during hunting season, making them harder to see and shoot, and if you do make a kill, you have less time to process it before the meat goes bad. But if you shoot a moose after 25 September, you are committing a crime.

The first time I had grasped the true extent of how at odds these two paradigms can be was in 2013. I was in Bethel, the regional hub of the Yukon-Kuskokwim delta, to cover the trial of twenty-three Yup'ik fishermen who had flouted the ban on fishing king salmon the previous summer. The courtroom was full, standing room only. Defendants, supporters, families hefting babies, a handful of journalists and cops. On the wall behind Judge Bruce Ward hung the Stars and Stripes and a traditional Yup'ik mask. Someone produced a Ziploc of salmon strips and passed it along the benches. Everyone took a piece and chewed on it, including the two state troopers. The courtroom began to stink like a fish market. The state called its first witness.

State Trooper Brett Scott Gibbens took the stand, spelt out his name, swore upon the Bible. He had a drooping, sandy moustache, a kindly yet sombre, doleful face, and standing upright behind the witness stand in his uniform and stab-proof vest he looked incapable of telling anything other than the truth. He began to recount how in June of last year, with the fishery ban on the king in place, he had been posted to patrol the rivers around Bethel. On 20 June he had come down the Kuskokwim River towards Napakiak to find a small fleet of boats, somewhere between twelve and sixteen of them, with what looked to be long, perhaps fifty-fathom, that is to say illegal, gill nets deployed. Many of the fishermen pulled their gear and left as he was identifying and rounding up the others. And yet getting caught was not an accident. The fishermen had press-released their intentions before setting out. The defence they were now mounting claimed that their right to catch the king salmon was protected, because it was religious. First amendment.

In 1979 the Alaska Supreme Court had heard a case in which Carlos Frank, an Athabascan native, had killed a moose out of season to provide for a funeral potlatch. He had defended his actions as constitutionally protected under the free exercise of religious beliefs. Moose meat was essential for the ritual of the funeral. The court laid out a two-part test to determine when a state was justified in violating an individual's religious rights. The individual had to satisfy three requirements: that a religion was involved; that the conduct in question was religiously based; and that the claimant was sincere. Those stipulations being met, it then fell to the state to prove that it had a compelling interest which justified its intervention. Frank was acquitted.

Felix Flynn was the first of the fishermen to take the stand. His Yup'ik name, he told us, translated as 'Journey in a Storm'.

'Is it okay if he occasionally breaks into Yup'ik?' asked Jim Davis, his lawyer, running a hand through a sweep of hair.

'We'll cross that bridge when we come to it,' said the judge.

Davis is one of the founders of the Northern Justice Project and he had taken this case pro bono. 'I used to fish with the elders and so I know how critical this is,' he told me. 'Telling a Yup'ik person he or she can't go out and fish is like telling a Catholic they can have mass, but only with the Eucharist or the wine, not both. If it was white people, Protestantism, Catholicism, Amish, then people would be bending over backwards. So why not the Yup'ik? Here it's like, hey, they had a good culture but I guess it's over now.'

Felix raised his hand and was sworn in. A short man with drooping moustaches and cheeks scarred with frostbite, he began by telling the court how when he was a boy his father would take him out herring fishing.

'To start with all I see is ocean,' he said. 'Then after a while there's glassy water, and there's other water that's not glassy. And that means the herring are here. That's what I learnt from my father.'

He had a way of drifting before the ends of his sentences, of looking about himself as though recently made blind.

'I'm subsistence. I was born and raised an Eskimo. It's in my blood. It's in my family blood.'

'And what does that mean to you, subsistence?' said Davis, leaning over with his hands spread on the podium.

'Subsistence is living from the land,' said Felix. 'It's what we've always done. We go hunt ducks and seals in the ocean in the springtime. Ptarmigan. Salmon. My grandfather and great grandfather told us we have to be very careful what we catch in the ocean. I'm a Catholic. The higher power created our earth and the food He provides for us. God made them for everyone. I was living subsistence even when I was in the military. My whole life. I make a fish camp every year, set net, and dry thirty, forty kings. I set a net last summer but there was too much fishing closure. Things have been rough.'

'And how did it feel not to be able to catch enough?' Davis asked him, his voice edged with compassion, glancing around at the gathered court to make sure that we all get this.

'I have a grandchild, two years old.' Felix paused and rubbed at his eyes. He was beginning to cry. So were several of the men in the gallery. 'My grandson said to me, when we gonna go check the net? And I couldn't say anything. We didn't have no net out then, and he'd seen me set it. And that hurt me inside. That's really painful.'

Michael Cresswell, the state trooper sitting next to me, leant over and whispered in my ear:

'This is momentous. This is climate change on trial.'

Cresswell had arrested half the men in the room. In a community of several thousand folk I imagined that could be awkward. These are men he'd see for the rest of his life, in the street, at the supermarket, at his kids' school. I asked him if there were any hard feelings. He smiled and shook his head.

'I have a real feeling that policy needs to shift with climate change,' he said. 'Civil disobedience has a long and illustrious history in the US. We're here to enforce the law and let them hopefully make new policy.'

'It depends on whose laws or whose supposed laws we were breaking,' Samuel Jackson of Akiak said to the judge. 'All of our traditional indicators were there. It's only common sense there were fish in the river. With all due respect, these elders behind me are the real biologists out there.'

Later, Ivan Ivan, the village chief of Akiak, would encapsulate for me the crux of the fishermen's defence. Chewing on tobacco, he told me the story of the muskrat, a small beaver-like creature once abundant round his village. A pelt brought a buck, thirty-two pelts made a coat. Need ten dollars? Go get ten muskrat.

'We used to trap the hell out of 'em,' Ivan said.

I could see where this was all heading. It was going to be a lesson that the Indigenous people knew how to protect their animals, that resources were infinite when well managed, that if they could do it with the muskrat then they could also do it with the fish.

'And now there's no more muskrat,' said Ivan.

I was taken aback. 'What happened?'

'We stopped hunting them. Social welfare meant people stopped trapping muskrat. It just didn't pay any more. What pays is food stamps. People stopped hunting the muskrat, and so the muskrat went away.'

In a paper submitted to the court preceding the trial, Jim Davis asserted: 'If Yup'ik people do not fish for king salmon, the king salmon spirit will be offended and it will not return to the river.' A second document continues:

> At the heart of the Yup'ik religious understanding of hunting and fishing is the notion that both animals and humans engaged in a 'collaborative reciprocity by which the animals *gave themselves* to the hunter in response to the hunter's respectful treatment of them as nonhuman persons' (citing Ann Fienup-Riordan). Thus the successful Yup'ik fisherman is not a person of great skill, but a person who obeys the traditional laws regarding the treatment of the salmon and who shows the salmon proper respect.

That is to say, if the salmon returns then it wants to be caught, and to snub this self-sacrifice is to offend its spirit and upset a natural balance. As such, any fishing ban is counterintuitive.

The document continues:

> A Yup'ik fisherman who is a sincere believer in his religious role as a steward of nature believes that he must fulfil his prescribed role to maintain this 'collaborative reciprocity' between hunter and game. Completely barring him from the salmon fishery thwarts the practice of a real religious belief. Under Yup'ik religious belief, this cycle of interplay between humans and animals helped perpetuate the seasons; without the maintaining of that balance, a new year will not follow the old one.

And yet now the seasons are out of balance, the distinction between the new year and the old one is a blur. Noah Okoviak, sixty-six years old, took the stand. He spoke through an interpreter, who began to cry halfway through his statement.

'Nobody here knows the weather,' Noah said. 'Nobody here knows how many fish will come. Only the Creator.'

'Do you think that king salmon have spirits?' asked the judge.

'Yes,' Noah said, unequivocally. 'All the species that we gather have spirits.'

When it was time for him to be sentenced every Yup'ik in the galley got to their feet, standing to show their respect for this elder, their hands clasped in front of them. Noah thanked the state troopers.

'They do their work for our Creator in trying to make a positive lifestyle,' he says. 'Only God can know how this situation will resolve.'

As with the others, Judge Ward found Noah sincere in his beliefs, yet guilty.

'The court finds,' he said, 'there is a compelling state interest in maintaining a viable, large and healthy salmon stock.'

Yet before moving on to the next defendant, the judge added something else.

'When this case goes up for appeal,' he said, 'the cold transcript will not reflect that everyone in the courtroom was standing. And that record will not reflect that there are a number of people in the courtroom with tears in their eyes.'

A mile down from the sonar, up the hill in Pilot Station, John Tinker sits, forking at moose liver and blueberry pancakes, watching his wife watch the Olympics. It's the swimming, Michael Phelps. It's Rio, and he's after his fifth gold, or his sixth. Beverly lies on the couch in front of it, their daughters ranging from toddlers to teenagers, spread out around the room. It's still the summer vacation, and they've been driving him batshit all morning. 'This man,' says the commentator of Phelps, 'carries all America's hopes on his broad shoulders.' John's hopes, by implication. Rio is very far away.

It's late in the morning, he's stiff as hell. He couldn't get going today. He's not getting any younger. He is going grey in his beard, but his shaved hair is still jet black. He has started wearing glasses, and they give him an owlish aspect. They hadn't got in till after midnight. A late opener, announced so late that Boreal, the buyers, hadn't got their shit together and sent any boats out for it, so they'd all had to go down to the factory after, waiting in line for hours to get paid. Driving back in the dark, no lights, bouncing through the waves, skylighting the hills to try and make out where he was. He'd come real close to a couple of logs. Pushing his luck again. You can only push your luck so far. His neighbour hadn't got in till five, and had scarcely anything to show for it. That was just how the cards fell sometimes.

From here down to the mouth was the only section of the Yukon where there was still a commercial fishery. There were silvers here, and the chum were still of sufficient quality. Although now that they had to avoid kings at all costs, where they could fish and when was much restricted: often the openers came only at the dog-end of the season once all the kings had passed. John had heard the complaints of the Athabascans upriver, that the Eskimos caught all their fish, that the commercial fishing had run the kings into the ground, but it wasn't like that. Down here had been just as bad. And up there they had jobs to go to. Down here there was nothing. There had to be a way of making some money, to pay for the boats and pay for the gas, and to pay for the nets that seem to have to change their mesh size every year. Commercial and subsistence are two sides of the same coin. One enables the other, and using the commercial to make the money to catch your own fish was little different to using whatever resources came to hand ten thousand years ago to make your net. At the end of the day, it's all just fishing.

John's house is a single-storey, rectangular prefab, like every other house in Pilot. There is a big wood stove in one corner, and they fire it with driftwood that they snag from out the river, trees from hundreds of miles distant; the spindly willows about the village give off no heat at all. The walls of the kitchen are plastered with the product of the journeys of eight kids through the school. Certificates for best pupil, best attendance, paintings of fishing and snowmachines and basketball. He and Beverly holding hands beneath a rainbow, *I love you, Mum and Dad*. The other day Sonny brought home *Spot*, the book they were reading in class. Sonny couldn't understand why anyone would want such a useless dog, let alone write a book about it. Spot couldn't pull a sled, or fight a bear off. He'd never done a day's work in his life.

John has a boy and seven girls. His two eldest were off having kids. One of the girls had moved to Anchorage, but the rest of them are here still. He hates going to visit her. You can't breathe, all the diesel and deep fry and perfume and sweat. You can't piss anywhere you want. But it seemed to hold some attraction for them. The biggest native village in Alaska, that was Anchorage. Full of Yup'iks like themselves, full of Athabascans, Tlingits, Aleuts, Inupiaq. Now his sister, Mary Jackson, had left too, off to run the AC store in Fort Yukon. He knew how homesick she was. He'd tried moving too, once, him and Beverly. He'd gone to Sitka, for seminary, and then to Kodiak, for business school. He'd lasted a year. He'd come home, like they all told him he would. He'd failed. Like any native Alaskan from a dry town that moves to a town with a liquor store, he thinks. Young and dumb.

You move us Yup'ik to the cities, he's always said, we fail. We might have wanted to move, try and make something of ourselves, but when we get there we fall down. We get addicted. He has seen it happen time and again. You make your choice, you swap wild grass for concrete, when you fall down you get hurt. There are four corners to Pilot Station, the place makes sense. There's everything you need here. Last year the school took a trip Outside, to California. One of his girls went. She saw a horse; it looked like a moose. She came back saying how in California, everything fun you have to pay for.

He knew he was supposed to want college for his kids. But Pilot Station wasn't a bad place to be, if you gotta be somewhere.

No one seemed to think any more that being a fisherman, or running a trapline, or raising a family that didn't drink and looked after one another, was anything to be proud of. You could do Native Studies now, at the University of Fairbanks. You could move to Fairbanks and study what you had left behind and get a degree from a white man when you finished. He wants his kids to grow up fishing, speaking Yup'ik, far off from all the violence and addiction of the cities. His parents never spoke Yup'ik to him. They said English was the future. He knows a few words, not much. It's compulsory in the schools now. Now all he needed was for one of his daughters to train as an engineer so that he might get some help with his truck.

All of a sudden, the CB radio whines and explodes into life. It sits in the corner, always on, buried beneath unwashed plates and shotgun shells, muttering awake from first thing in the morning, and chattering away to itself until bedtime, the community's subconscious.

'Good morning, Pilot Station, good morning.' It is Lindsay, from the sonar station. 'Commercial fishing will be open for one period in District 2 on Friday, August 26 from 1 p.m. until 9 p.m. This will be an eight-hour commercial fishing period with gillnets restricted to six-inch or smaller mesh size.'

Beverly looks up from the TV. John rubs a hand across his face.

'Aw shit,' he says.

There is quiet, the hum of static.

'This has been an announcement by the Alaska Department of Fish and Game in consultation with the U.S. Fish and Wildlife Service.'

'Thanks, Lindsay,' comes another voice from the radio.

'Superstar,' says someone else.

Michael Phelps has won another gold.

'Shit,' John says, again. He folds the last pancake and shoves it in whole and swallows down his coffee. 'Sonny!' he shouts. 'Where you at?'

Sonny, eight years old, his only son in a house of women, totters out of the bedroom in a pair of his mother's high heels, the sequinned silver ones.

John stares at him. 'What the hell are you wearing, son?' he says.

'Sonny's got girls' socks,' says one of the girls. 'And he likes girls' pants and girls' skirts.'

Sonny nods at the truth of it.

'Go and take that shit off,' John says. 'You're coming fishing.'

He texts his deckhand. He fills his thermos with coffee from the machine, and kisses Beverly goodbye. He goes out into the *qanisaq*, the Arctic porch, pulls on his Helly Hansens, shouts for Sonny, and goes down and starts the truck.

He lights a cigarette and cracks the window. He's tried to quit he doesn't know how many times. The weather is still grey and cold and damp, the light held in a fist. Trucks and four-wheelers buzz back and forth. Everyone is getting ready, the village has come awake. He sees Terry, coming the other way, and pulls to a stop alongside him.

'Hey Boyer,' Terry says.

'You heard the opener?' John says. 'One to nine.'

'Yeah, I heard,' says Terry. 'I can't do another, not today. I'm not young any more.'

'Lots of fish!' says John. 'Lots of fish!'

'Not like there used to be.' Terry peers across at Sonny, sitting in the front seat, legs stuck out in front of him. 'Your Dad pay you?' he says.

Sonny nods.

'I should hope so,' he says. 'How much you get?'

'Fifteen bucks.'

'That's a lot of money.'

'I'm going to save it until I got twenty thousand bucks and

then I'm going to buy my Dad a three-hundred-horse Yamaha,' he says.

Terry raises his eyebrows. 'That won't last,' he says. 'Be safe. There's a storm coming in.'

'He's not a real fisherman,' John says to Sonny as they drive down to the docks. 'Real fishermen go fishing.'

They swing by the store to get supplies, some candy for Sonny, some cigarettes, and then they drive down to the dock. There is a single dead humpy, sloshing about in the shallows like an empty sack. Sonny pokes at it with his foot. Walter, his deckhand, is already there, thumbing through his phone. *Your boat's ugly and your bait smells*, it says across his hoodie. The boat is full of trash and blood, Pop Tart wrappers and Capri Suns, the tote still full of gore. He hates leaving it in such a mess, it gets real stink, but they were beat when they got in. At least they'd had a good day of it. He'd put the picture on Facebook of their six containers, more than four hundred fish, overflowing onto the deck. Mary, his sister up in Fort Yukon, had liked it straightaway. She was so far from home, that girl.

Three thousand pounds of fish yesterday, mostly coho and some chum: $2,400. Not bad for a day's work, though once he'd paid off Walter and paid for gas, it wasn't much with ten mouths to feed. It wasn't like it used to be. Back when they let them fish for kings, that was the good money. Once he made $14,000 in a single opener. More money than sense, they'd used to joke. But that was years ago.

The wind is blowing the waves up into whitecaps. The wind is always especially swift through here, buffeting over the last of the bluffs, the last land higher than a blueberry bush before you get to Russia. It's not the sort of day he'd normally go out on, but you have to fish the openers when you get them.

'Get your float coat on, Sonny,' he says to him, and he pulls on his own lifejacket, buckling it up, thin and beat, more used for kneeling on than wearing. John steers an arc out into the middle of the flow and turns the skiff to face the current. He gives the wheel to Sonny, who takes it, face fixed serious with the gravity of the task, and he and Walter heft the tote of slops onto the gunwale and tip it overboard. The seagulls are massing already. He gathers the trash into a sack and scrubs the blood from off the deck and from off the bloody floats.

John takes the wheel from Sonny and whoops, and he opens up the throttle. 'Hold on, boys!' he shouts. The bow bucks and rears and the skiff wheels away upriver, hammering through the waves. A mob of whitefronts bursts from shallows. Above the engine they cannot hear each other. Sonny huddles down inside his coat. Walter sits looking backward, smoking a cigarette, watching the wake pan out, watching other skiffs peeling away from the docks like bats leaving the roost, heading for fishing spots that they have chosen by a combination of logic and divination.

The river flexes about like a snake. Far distant they can see the squalls of rain coming. Clouds, everywhere, white fading to black, and a low-hung mist between. A whole world of weather, an Arctic sky. It is good to be fishing again. It's good that Fish and Game are letting them. There's nothing worse than sitting around at home with all the fish swimming past and nothing you can do about it.

There were those in town who agreed with what those fishermen on the Kuskokwim had done with their court case. The Yup'ik didn't like being told what to do. But there were others, like his father-in-law, who had gotten mad about it. Management was there for a reason, they said. They couldn't just think about the short term, and all of this spiritual crap. 'Some people just don't want to follow the rules,' his father-in-law had said. Now you even got people, politicians even, trying to explain away all the social problems of the Yup'ik because of a lack of kings. John had kids. He had to think about that. And if he wanted Sonny to grow up a fisherman, well perhaps they'd just have to do what the scientists told them to for a while. Even if they were all white.

They reach the north tip of Spangle Island an hour later. There are a couple of other boats out here with their nets already in, but not as many as he'd expected. Walter lobs the orange buoy into the river and pays out the floats over the bow, the net trailing in behind, as John reverses the skiff in a gentle curve across the water and holds it there, the engine idling. Once the net was a hundred feet, but it was probably down to eighty by now, with all the repairs they've made. A five and seven-eighths inch mesh, just right for those female silvers. An eighth of an inch can make all the difference.

They settle in to wait. The rain is coming down and beading on their oilskins. He can feel the shift in the season, it has toppled into autumn. He pours himself a coffee, sweet and milky, from the thermos, and lights a cigarette and sits there, looking about himself. He pulls out his phone. The background is of Sonny, holding a spruce chicken by the scruff. He shows it to Walter.

'He killed that with a rock,' John says.

He sips at his coffee. He is looking out across the water when the river explodes into rapids of turbulence, as though caught in a downpour. A piece of muscle blazing silver, flexing, flapping like a flag in a stiff wind, skimming about the surface. The power of the fish was only apparent when held like this, gilled, a storm kept in a bottle. He had seen it often in his forty years, and yet it moved him still. Yanked from a course as magnetic as an orbit, thrashing with the force of its unrealized biology, an imperative that had sent it on a journey of five years, that had led it away from this place only to eventually drive it back, and the trawlers and fish wheels, the bears and logjams and sea lions and diseases it had dodged up until now. Did they give themselves to be taken willingly? Well maybe they did, but that made the sacrifice no smaller. After all, he thinks, even Jesus had struggled on the cross.

John stamps on the floor of the skiff. When there's a pulse coming through it helps to scare them up and get them tangled. They see another hit, and another. He smiles. He'd judged it right. He knew they'd be here. He just knew. *Neqpik*, they call salmon. In its literal translation, it means 'real food'. One of the few words he knew.

'If we don't get that net out soon,' says Walter, 'we're not going to be able to haul it.'

They bend to it, the three of them. His son standing close behind him, grunting like the men. This made it worth it. All the hard work, and all the cold, and all the waiting. They haul it, hand over hand, the nylon biting into their gloves. Sonny pressed warm against John's legs. The salmon writhe, gleaming in this dull light. Eight, nine, eleven, he counts them in. Walter pries them from the mesh and drops them to the tote, and Sonny stands, peering in, mesmerized. It will be a good day, if this keeps up. They have come back, just as they should, just as he knew they would, to sow the seeds for the next generation.

September is here, and the weather has changed. This morning we woke and the sky was dark, and at dawn the leaves on the willows had transformed to a jaundiced, liver-speckled yellow. The cranes have turned to white. A single skiff is indeterminably distant, floating in the mist like some ancient Chinese painting, a single man bowed to his net. And the geese are going south. All day, thousands of them overhead, all keyed to the same compass point, strung out in simple geometries, lines and arches and chevrons of them, a canopy of geese. It makes me nervous to still be on the river if the geese are on the move. By the time the weather hits, when their breeding pools are under several feet of ice, they will be half a hemisphere away, bobbing in the sunshine, feeding on southern grain. They pass overhead, and we watch them, wave upon wave of them, their cacophonies eventually dim and distant as they fade into the southeast, and summer is like some tangible thing that I can see disappearing along with them.

We are so close to the sea now. We stop for lunch and walk out on the tundra, picking blueberries. The moult feathers of the geese vibrate, snagged on the grasses. There are wizened mushrooms, well past their best. This is a landscape that we have not yet seen, now so far out on the delta. Underfoot is a miniature tangle of growth, of exquisite lichens and sedges, mosses and flowers. There are squat bonsai trees two inches high that may be fifty years old, their spread stunted by the permafrost that entombs their tiny roots. The land has the feel of an Alpine meadow, rolling out across the plateau to the blue hills in the distance. You could run for minutes with your eyes closed. The horizon draws your eye in, tugs at the feet. What would it be like, I wonder, to just start walking? No fences, no roads, for many hundreds of miles. Wilderness, true wilderness. The word runs like a mantra, whatever it might mean. When would you next meet a person? The loudest thing is my own body, digesting. And a noise in my ears, which is maybe my own blood, or maybe the silence that is the background hum of the universe. In the distance, more squalls of geese.

We push on. Three months ago the kings would have come up through here. The weather gallops toward autumn. We drive the canoe over breaking waves and water swills into the boat. In the stern, I bail. It rains and rains. There is little meaningful distinction between the river and the sky. It is becoming hard to remember a time when we did not get up and paddle every day. But in a couple of days' time we will be at the journey's end. As the river braids further there are long sloughs that we take to keep out of the worst of the weather. We camp on muddy banks that are covered with yellowed, dying grasses. At night the wood is too wet for fires, and we eat Pilot crackers spread with peanut butter and jam. In the morning we pull on wet gloves and wet shoes, half-frozen, and we paddle on again.

The first sign of Emmonak, the last village before the ocean, is of four white wind turbines poking complicatedly up above the horizon's line. It is the first piece of infrastructure that I have seen since we passed beneath the bridge two months ago. I had forgotten that humanity was capable of such domination of a landscape. The turbines revolve languidly, and the shape and the size and the colour of them are out of all coherence with the rest. 'You wanna hide something out here,' someone once told me, 'don't put it in anything white. Apart from the heads of bald eagles, there ain't nothing white out here.'

All day long we have seen fishermen, out drifting nets in the channels. They race past us, oilskinned, sunglassed, a hand held high in salute, and we turn to face the slop of their wakes before they flip us broadside. There is much more traffic here, more than we have seen anywhere since the start. But for one small plant near Pilot Station, Emmonak has the only commercial fishery on the Yukon, and the only factory of any decent size. There are enough chum and silver this close to the ocean to still make a go at a business. They used to fish for king too, of course, but that was years ago. Tenders – the large boats that gather up the fishermen's catch and transport it, on ice, back to the factory – move amongst the skiffs, and as they turn about for home we follow on behind. They peel off the main stem of the river, into Kwiguk Arm, heading for the processing plant. Once this would have seemed like a big river, but leaving behind the enormity of the Yukon it feels more of a creek, not much wider than the Thames. Rusted machinery lines the north bank, rusted shipping containers are stacked along the wharves, a rusting, antique barge is moored to two antique bulldozers. Construction has slipped where the permafrost has thawed; it looks like the flotsam of a storm. Ahead we see warehouses and factories and cranes. Seagulls flock the sky. I have seen so little on this scale for so long I feel as though we are docking in Seattle.

What must they think of our canoe? Like taking a horse and cart down a motorway. We turn sharply into the oncoming wakes from the tenders, bucking over them, nothing for it but to set the paddles wide to keep us stable as best we can. In the calmer moments we edge closer to the port. This is suddenly terrifying. Five minutes from home would be quite a place to capsize, and frankly surrounded by these vast barges and by people trying to make a living, I feel like we're asking for it. At last, nipping between two skiffs, we slip into the harbour and moor up on a pontoon. We sit there, calming down. We've made it.

I get out to stretch my legs. There are little kids hanging around the moorings, their faces grubby and gnat-bitten, one with 'BEAST' written across his singlet, and as we unload our mound of things they crowd about us, their hands stuck out for sweets and money.

'C'mon, give us a buck.'

'What do you want money for?' I say.

'So I'm not always broke,' he says, like he's heard his Dad say it, and hawks some spit.

The fish processing plant in Emmonak is run by a company called Kwik'pak, the Yup'ik word for the Yukon, and their buildings are just up from where we have moored. I go into their office to ask what to do with the canoe, and they ask us if we would like to stay in their bunkhouse. We are soaked through, and we move in directly. We have a room to ourselves, two single beds with a desk between them, lights above the beds for reading. We stash the canoe beneath the window outside.

Kwik'pak has luxuries to which we have grown unaccustomed. There are hot showers; there is as much cereal as we can eat. We wash our clothes in a washing machine. For dinner there is roast beef and mash, horseradish and green beans. After, we sit in the canteen drinking coffee after coffee. As we dry off and warm up I become more aware of how exhausted everyone else is. People look shattered, the walking dead, great bags beneath their eyes. We are an object of faint curiosity, no more. The man across from me pushes his plate back, pats his belly, yawns.

'Long day?' I say.

'If you ask for time off,' he says, 'they call you a motherfucker.'

They eat with their heads down, in silence, staring at their plates, and when done they get up and walk out.

After dinner we go for a stroll. Outside, in rubber boots, we make our way through the mud and gore that appear to be the very foundations of the factory, a miasma of filth and fish blood and fish guts hanging on the air. Kids walk through the mud, hoods up, biceps out, wrung out with lack of sleep. A bobcat purrs past us, leaving ruts in the mire. Right now, I can't face speaking to anyone about fish. We go to bed, and lie there, listening to the factory through the night.

We spend some days walking around town. There is nowhere we have to be. It is odd: this place that we have spent so long in getting to, that I spent so many months looking at on maps at home in London, and so many more while sitting in the stern of the canoe, is a place where other people simply live and go about their business. We sit in the Family Restaurant, eating sandwiches, playing pool. One day we go to the sports hall, where there is supposed to be a demonstration of traditional dancing and fiddle music, but everyone has gone fishing. Emmonak looks much like the other villages at this end of the river, muddy and sprawling, but so close to the sea it feels yet more exposed and stark. There are no log cabins now – there are no logs – and everything is made from ply and corrugate. Buildings slump in varying states of disrepair, and the turbines revolve above the town like a vision of some future world.

In the seventies, all of the big outfits working out of Emmonak were Japanese. The biggest food company in Japan had offices here. The two-hundred-mile limit had recently come in, and the Japanese were no longer permitted to fish off the Alaskan coast. Japanese companies bought their way into American salmon packers as a way of securing supply. They only wanted chum, and so all the kings went to New York, dry salted and packed in tierces and sold to Jewish smokehouses to make lox. When aquaculture became established in the mid-eighties the New York smokers switched to farmed fish, and the Japanese were left with a king run on their hands and no idea what to do with it. This was before the fresh-fish market boomed, and besides, there was no refrigeration in Emmonak. The Japanese decided on king salmon flakes, for sprinkling on rice, and they sold them back home at a price not far off gold dust. Business was booming. And then, the runs collapsed.

Today it is only the native owned Kwik'pak that is trading out of town. Kwik'pak believe that the runs are picking up again, that the kings are looking healthier, fatter and stronger, and that the good times are not far away. They are the only seafood company, worldwide, to have been given Fair Trade status. One of the main tenets of the business is to improve the economic situation of these communities through reinvestment. There is little else here. Come the winter, outside of the fishing season, there are forty jobs for eight hundred residents.

And Emmonak is flourishing. Unlike Alaskan villages upriver, where the amenities are closing, here they have just built a new AC store, a new post office, a larger school. Nearly half the people here speak Yup'ik as their first language. When the Russians first arrived at the mouth of the Yukon, they could not figure out the maze of sandbars that confronted them, and rather than risk running aground they sailed north up the coast and portaged their boats over to the river that way. That means this area went untouched by white outsiders longer than most. It also means that the changes that have come here have been even more rapid than what the rest of Alaska has experienced. That brings with it its own particular demographics. This is the youngest county in the United States: 47 per cent of the population is under eighteen. It is the fourth poorest county in the United States, and it has the highest unemployment of *any* county: more than a quarter of the population lives below the poverty line. Compare this to the North Slope, 300 miles to the north, a census area that is broadly Inuit Eskimo, and which is the twentieth richest county in the whole of America.

The history of this disparity derives in part from the creation of the Alaska Native Claims Settlement Act (ANCSA).

Following the discovery of oil in Prudhoe Bay in 1968, before the pipeline could be built and oil leases sold, some decisions had to be made about what land was held by whom. Governor Walter Hickel reckoned that 'you can only claim title to land by conquest or purchase. Just because your granddaddy chased a moose across some property doesn't mean you own it.' Stewart Udall, Secretary of the Interior, disagreed, as did, unsurprisingly, the Alaskan Natives. Udall suspended all transactions on land until the dispute was settled. In 1971, ANCSA was ratified. It gave the natives one billion dollars and forty million acres of land (one-tenth of Alaska), divided along cultural and tribal lines into twelve regions. In each region a native corporation was established to manage the local wealth (with a thirteenth corporation for non-resident natives). Overnight the natives became stockholders, with the expectation of annual dividends. Much of the inherited land was rich in deposits of minerals, in oil and gas, timber and fish (although much wasn't, and this accounts for the vastly different living standards, seen in the corporations to this day). It was, to quote essayist John McPhee, 'widely described as the most open-handed and enlightened piece of legislation that has ever dealt with aboriginal people'.

As a place to study the effects of capitalism in microcosm you could do worse than the Native Corporations of Alaska. The debates between the proponents of the free market and the anti-capitalists are manifest in the Indigenous villages: whether land should be owned by corporations or by the people who live on that land; whether the material benefits brought to peoples' lives justify the increased disparity in living standards; whether a system that maximizes the use of resources for profit makes sense when their maximization ultimately impoverishes the land they were extracted from; whether constant growth can be maintained in the face of natural limits. In Point Hope, on Alaska's North Slope, it is evident how the seismic equipment used for oil exploration is affecting the communication of marine mammals that the village depends upon for food, or what the consequences of a spill would be in its pristine environment. But, in a village of fewer than seven hundred people, there is a fire station, a health clinic, water and sewage treatment facilities, a power station and a high school. Every desk in school has a MacBook on it. There are several millionaires. All of this is built on oil money, from contracts sold by the Arctic Slope Regional Corporation.

One evening, after dinner, I walk across the bog outside to see Jim Friedman. I find him in his office, a shack attached to the side of a larger Portakabin. He is slumped in a swivel chair, surrounded by a crush of teenage girls who are working their way through packets of Oreos. Scrawled in marker across the inside of the office door: *Don't let a dead fish beat ya.* I wedge myself into one corner.

'Cup of tea?' says Jim.

He scoops a mess of papers and wrappers to one side, and there beneath all of it is a kettle. He fills it with water from a jug, drops in two bags of Earl Grey and two bags of raspberry leaf, and flicks it on to boil.

'I've never seen anyone make tea in the kettle,' I say.

'Where do you make it?' he says.

Jim is of indeterminate age, perhaps in his late fifties. Every day I see him he is dressed in overalls and a beanie made of muskox hair that has been darkened several shades by his sweat along the brim. He moved to Alaska with a job at the Galena Airforce Base.

'A guy said to me on the plane on the way up, "Don't stay for more than one winter. If you do you'll be hooked." Well,' says Jim, 'he was right.'

He hands me my cup of tea.

'I've been in the egg business for thirty years. The Egg Man, that's what they call me. I was sitting in my office one time and that song came on the radio. I am the walrus, I am the egg man. I walked around smiling for a week after that. John Lennon wrote that, about his buddy Eric Burden. He was a real pervert. He used to crack chicken eggs onto his women. And John Lennon was reading that book by the guy that wrote *Alice in Wonderland*: I am the walrus, I am the carpenter. And they came up with that together when they were on LSD. True story.' He grins his sloppy grin. 'Coo coo ka choo,' he says.

The Egg Man smells of eggs, profoundly. Not only his clothes, but his entire self. I come across him one evening in the showers, and he is standing there, red and steaming, rubbing himself with a threadbare towel, and the smell is in no way diminished. Whether it is his workplace or his diet that has imbued him is unclear. Jim eats fish eggs three times daily. Jim swears by fish eggs. He says he can remember phone numbers by heart, no need for pen and paper.

'You remember when Gary Kasparov lost to Big Blue at chess?' he says. 'He went away and he ate fish eggs. He ran and he ate fish eggs. He lifted weights and he ate fish eggs. And then he came back and he beat Big Blue. That was his secret. Fish eggs. The Russians know. Here, have a cookie.'

I follow Jim next door, onto the egg factory floor. It is fridge cold. He walks through the place, hands in the pockets of his overalls, beaming, as though the world is moving to clockwork of his own devising. The kids can start here before they reach eighteen, because there are no knives involved. Later they will graduate to a spot in the fish-processing factory, and from there, perhaps, to the boats. They have been at school all day. To the sounds of Coolio, Eminem, Notorious B.I.G., the eggs move from station to station.

'I try to get them to listen to proper music,' Jim shouts. 'The Who, Bob Dylan. All they like is rap.'

The full skeins of eggs come in from the processing barge, hundreds to a bucket. The workers burst the sacs open, and force them through a mesh to separate each egg from its membrane. From there they are transferred to a brine with the same salt-water ratio as the ocean ten miles downriver, a device like a top-loading washing machine that spins them in a saline foam, and then they are scooped out and left to drain. The girls huddle and gossip. They spray each other with the hose. The boys get the girls in headlocks because they love them.

'Freshness,' Jim intones. 'That's the secret to caviar. I make good caviar or I don't make it at all.'

I ask him which salmon makes the best eggs.

'The Russians like the coho. The Japanese like the chum. It's like wine. It's all good, just different tastes.'

This, then, is how caviar is made. It is made by Yup'ik teenagers in the evenings after school, in a bloody Portakabin in a muddy swamp, with Puff Daddy on the speakers. There are many trays of eggs, waiting for refrigeration, and he invites me to dive in. They drip from his fingers like pearls. There is something wonderfully decadent about eating a luxury product by the handful. This season they have so far made forty tonnes of caviar. Jim is aiming for fifty. If I wanted to I could sit here and eat till I was sick. I think Jim would let me. I think he might encourage it. And I would not forget a single moment.

One of the girls comes up to us. 'Can I work all night?' she says.

'How old are you?' says Jim.

'Seventeen.'

'No, then.'

'Why not?'

'You have to be eighteen,' he says. 'Go get a fake ID. Go put on make-up and a wig.'

'I could have blonde hair and blue eyes,' she giggles, and goes back to her station.

Back in his office there is a girl slumped at Jim's desk, howling over a lost boyfriend, and so we walk outside to talk. Jim produces a pack of cigarettes and rips the filter from the end of one and lights it.

'So you've seen the river up and down,' I say. 'Who's right about managing the king?'

'It's like the argument between Yasser Arafat and Israel, the arguing over the fish between the Indians and the Eskimos.'

'But who do you think is right?' I say.

He pauses, considering.

'They both are,' he says. 'And there would be no problems managing it if the government stayed out of it.'

This is, of course, the response one would expect from private business. Deregulation, and no state intervention. But then the Canadian First Nations have been motivated to protect the fish themselves because they are not expecting anyone else to do it for them. Jim's whole approach, and why the kids here love him, is because he believes in them. They are from a place where no one – not government, not schoolteachers, not their parents – has trusted in their capabilities. A job, and a place to be safe, provides them with far more than a few bucks an hour.

'I was a mischief kid,' Jim says. 'Everybody was. That's why I understand what they're going through. You have to treat them with respect. How do you have hope when you live in such conditions? Half of these kids don't even have a flushing toilet. You got the most horrible conditions and the finest salmon in the world here. For me, if the fish are poor but I teach the kid a lesson, that's been worth it. The fish'll come back. A life won't.'

By the time I leave Jim it is dark, but there is no let-up in production. Halogen lights illuminate the forklifts that are zipping back and forth. The processing barge steams in their glow. Men stand around in small groups, smoking. The season is one immense day that stretches from June to the beginning of

September. They have just passed the mark of a million fish, a record, and the summer still has some days left to be wrung from it.

The tender is called *Nunataq*, painted on the side in great sloppy, glossy letters. She is moored alongside others of the fleet: *Qipngayagaq*, *Akuleraq*, *Quipngayak*, the names of local villages, and she is painted in the primary colours of a Caribbean fishing village, blue hull, blue crane, yellow doors, red walls, the table red and the window frames yellow, all in the same bright gloss, so that the interior of the wheelhouse seems to glow with radiance and hope. Pinned above the instrument panel is a photo of an ancient woman, a face like a well-thumbed book, wrapped in a headscarf and fur-lined parka, black berries of eyes, holding up a gallon bucket of salmonberries for the camera.

Lindsey, the tender's captain, stands at the instrument panel, sipping on a Red Bull that he rests back in a wire holder fixed to the wall, made during one interminable afternoon in the boat. Through the glass he can see Tyrone muddling around on deck, shifting ropes about. The day is grey and close. A blizzard of gulls crowd the water to the stern, alongside the processing barge, flocking and diving for the slops, the viscera and heads. Lindsey checks his Facebook while he waits. The engine drums, the cabin warming from the heat of its pipes that climb up the far wall. John walks through the door, a Nike bag slung across his back, and Lindsey greets him with a nod and some few words and looks back at his phone. John sets the bag on the table and takes out four cans of Red Bull and a carton of Marlboros and a packet of peach rings, and lines them all up on the table side by side as though he is about to sell them off. He opens the carton and peels the cellophane from a pack and begins to transfer the cigarettes, one after another, into a slim metal case that he keeps in his jeans pocket.

Tyrone knocks on the window and gives Lindsey the thumbs up. Lindsey nods back. Tyrone hoists the midline and throws it to the deck and jumps down behind it, his long thin frame landing in a half squat, a cigarette still clamped between his lips like a fishhook, smoke rising like breath in the chill of the morning. Lindsey steers out into the channel where the skiffs zip up and down. Tyrone stands and coils the rope in and busies himself about the deck, hauling up the fenders, making the totes fast. He walks to the door of the wheelhouse and he stands there, one forearm on each side of the frame so that he fills the door entirely, hood up, cigarette flopping from his mouth. There are shadows underneath his eyes. He has a bristle of moustache and beard, the hairs scant enough that you could count them. The peak of a wide-brimmed cap sticking out from beneath the hood of a camouflage sweater. They are all exhausted. Yesterday they brought in twenty-six totes of salmon. More than twenty-six, in fact, because by the end of the day the totes were all so full that they were overflowing into the hold, 46,000 pounds of fish. This was Lindsey's eleventh year on the boats, seven as deckhand and four as captain, and it was the biggest day he'd ever had. By the time they were through with offloading them it was gone three in the morning. A few hours sleep and back out the door. The last few days of the season, and Kwik'pak wasn't letting up.

'Hold's got stink real *bad*, bro,' says Tyrone through his cigarette.

'Go wash it out then,' says Lindsey, not taking his eyes from the depth gauge. This bit of the channel is always shifting and he's known captains plenty more experienced than him who have run aground through here.

'Aw, shit,' says Tyrone.

The boat moves out of Kwiguk Arm and into the main stem of the Yukon. Lindsey sets a course on the GPS and holds the wheel steady with a piece of string that is tied onto the dashboard. Nunam Iqua is a two-hour drive away. Lindsey can worry less about the sandbars now. He lights a cigarette and takes another sip of Red Bull, and stares out at the passing landscape. The low flat banks, some few sparse fish camps. A skiff zips past them. Out on deck, Tyrone flips a sideways V in greeting, like the scissors in paper-scissors-stone, and the two men both make the same gesture back, holding it for long seconds as they stand sentinel in their rain gear, until the skiff draws level and passes the tender, and their wake slaps up against the hull.

His eleventh season, Lindsey thinks, coming to an end. Soon he'll have to start looking for winter work. He might try in the school this year, as a janitor or something. Summertime in Emmonak there's a job for everyone who wants one, so much work that they come in from Mountain Village, Marshall, St Michael's, Hooper Bay even. Wintertime you have to fight tooth and claw to be allowed to scrub someone's toilet. Families split up over who got jobs. He remembers the days when he made enough money from fishing to sit on his ass and drink all winter. That was after the army, before the kids came along. He didn't even have enough time to catch his own fish, these days.

'What you doing when the season's done?' he says to John, who is still arranging his cigarettes.

John looks up. 'My boy's coming up from Mountain,' he says, speaking above the engine. 'His mum's off to college. I might take him travelling this winter. California or someplace. It would be nice to teach him a thing or two.'

John looks back down at the cigarette case, contemplates it, and snaps it shut. He is twenty-two. His boy is one and a half.

'My ex's family, they're good people,' he continues, following on some train of thought. 'They live real subsistence. I like that. They don't eat much of that processed food. They might go down to the store sometimes for some grub, you know, but just enough to go back out hunting for more subsistence.'

Tyrone comes in and flops down on the bench opposite John. John and Tyrone are cousins. Lindsey is married to John's sister. They're all family. Everyone is.

'Man,' Tyrone says. 'I leaned on one of those totes and now my jeans are fucking stinky.'

Lindsey misses the army. He had been in for four years. He'd travelled. He'd been to Iraq on a six-month tour. Who ever thought a kid from Emmonak would get out to the desert? Out on the convoys and the patrols. He'd seen some pretty fucked-up shit. Some of the other guys, it messed them up. They got that PTSD. Not him. His Dad was a Vietnam vet. Now there was a real messed-up war for you. His Dad still wasn't right. He couldn't stand his kid being off on tour, it wasn't good for his heart. And when it came time to re-enlist his Dad gave him hell for it, and so he didn't. He missed it. His Dad had encouraged him to get treatment, for the PTSD, but he kept telling him he didn't need it. He'd learnt how to treat himself. He'd learnt how to not think about it. Just keep busy, boy. Keep on doing things. Sometimes he'd feel it coming on, and he'd think, fuck, best get busy. It worked. He got a lot done. Maybe he didn't miss those winters of drinking any more.

Two hours later, they drop anchor off an island a couple of miles from Nunam Iqua. 'The end of the tundra', it means, and it is. The village sits where the river meets the sea, on a small and eroding point. No roads, just boardwalks. Two hundred people. Lindsey's heard people say that in 1969, when Neil Armstrong landed on the moon, they filmed it out at Nunam Iqua, in the winter. That the Stars and Stripes is still on a pole out there, somewhere. He can believe it. He kills the engine. The boat sighs and is still. He looks out of the window, trying to make out the far shore. From one bank to the other, here, the river is seven miles across.

They listen to the water's lap. A few geese rush past overhead. Lindsey gets down in the engine bay and starts to change an air filter. The wind blows round the boat. Tyrone opens two bottles of water from a crate of forty-eight and pours them one after another into the coffee machine and turns it on, without coffee. When it brews he pours the water over a disposable bowl of instant noodles, adds the packet of flavour, and reseals the bowl so it can steam. After a few minutes he reopens it and pokes at it with a fork.

'Native food,' Tyrone says, grinning to himself. He slurps it down. When he's done he sets the plastic fork down and sits there contemplating his hand as though it is an object he is holding.

'Look at this *hand*, man,' he says. 'All crunchy.'

'Why don't you go to clinic?' says John.

'If it's broke they won't let me work. You remember when I bust my shoulder? That was when Rick took my spot on unc's boat. That shoulder was *nasty*, man. Popped right out. I grabbed that tote of fish and I looked the other way and I just started walking. That was pretty cool but messed up same time.'

'Three hours' sleep,' mutters John.

'I didn't sleep at all, bro,' says Tyrone. 'I was banging the girl next door. I got them both, now. Both the sisters.'

'I thought Josie was with Billy.'

Tyrone grins widely. 'I made them split up,' he says.

'You're mischief,' Lindsey shouts from down in the engine bay, his voice echoing about.

A skiff pulls up alongside and ties a line. An old man, dressed from head to toe in orange oilskins, waves up at them. There is a dead seal slumped in the back of his boat, alongside a harpoon. Tyrone goes out and sits in the crane that is fixed to the front of the deck, a cigarette stuck in his mouth. He swings out the arm above the skiff and lowers the bucket, and the man throws in the three fish from his tote. Tyrone then raises the arm until the bucket swings free.

'Fourteen pounds,' he shouts, and then mimes it through the window to Lindsey, a single finger and then four. The man climbs the few rails to the cabin and pushes through the door, radiating cold.

'Hey Al P,' says John, looking up from his phone.

'Damn,' says Al P. 'What a day.'

'How you doing out there?'

'Catching more sticks than fish,' Al says. 'Never had a season so bad. My partner just went to college. Anchorage auto-mechanics. So I got no help. I need help in this wind. The current's real strong.'

'The wind's from the north,' says Lindsey. 'Not good for the fish. I seen a couple people pack it in already.'

'I wanted to fish on the bigger tide tonight,' says Al. 'But they won't let us have that one.'

'Fourteen bucks for the cohos.' Lindsey hands him the paper-work. The prices are chalked on a board. A dollar for a pound of coho, seventy-five cents for chum.

'How much is a quart of your Yama lube?'

'Eleven bucks.'

'Shit,' Al says. He takes off his glasses and rubs at his eyes, his cheeks webbed with scars of frostbite. 'Give me one, then.'

He stands there, vulnerable and blinking, and then replaces his glasses and pockets his three dollars.

'Well, I guess I'll go and try out there, then.'

He looks around the cabin. 'Warm in here,' he says. 'You haven't got a cigarette, have you?'

John takes out his case and hands him one.

'Well, quyana, you guys.' Al pockets two 7 Ups when Lind-sey's back is turned and hustles out. They watch him go, and shut the door on the cold.

The opening closes at six, and soon after the boats start arriving, two or three of them at a time. For many it is the last day of the season. There's a few days of turnaround to get the boats fixed up before moose hunting begins. There are men by themselves; two men; families and dogs; young teenagers cold and miserable, jabbing at their phones. Tyrone weighs out the catches. Some people have done well today, and others have done badly, and will have spent far more on gas than what they've made. People fishing just metres apart can make vastly different fortunes. The ever present possibility of a windfall gives the fisherman a fundamentally different outlook from the farmer.

It is gone seven, and they are waiting on the last of the boats when news comes in on the radio that two boats have run aground coming up from the Black River.

'Ah, sucks,' says Tyrone. 'Gotta wait.'

Lindsey shuts the engine off and stares out at the water. He picks up his phone and walks out on deck to break the bad news to his wife. Tyrone pulls off his hoodie and T-shirt and lies down, topless, on the bottom bunk.

'If you ain't caught up with thinking,' says Tyrone, 'this is a good place to make you think.'

'If only we got paid hourly,' says John.

Lindsey comes back in. The clock ticks.

'We shoulda brought out your tattoo gun,' John says.

They all stare at their phones, the glare lighting their faces as the day fades. The last two boats don't show until after ten, after running aground twice more in the labyrinthine shallows.

It is darkening now, the sky pulling out into pinks and blues, the light dying. Tyrone stands on the back deck, looking out at the wake unfurling, smoking. Skiffs race past them, and he wishes he was in one and back home in half an hour with his buddies, instead of the two-hour journey, and then the unloading, and the hose-down and the maintenance. A juvenile gull keeps pace with them, riding in their slipstream. There is so little in the landscape that anything is notable. A cluster of skiffs is gathered together in the middle channel and he peers to make out what they are up to, but with the distance and the light he cannot tell. Maybe they got a seal, or a beluga. The gull peels off, climbing, heading west, and Tyrone shivers with the coming night and flicks his cigarette to the river in a shower of sparks and steps back inside the cabin.

It is dark now, and they are silent, the day done with, in companionship. A dull thud of dance music on the speakers. Lindsey dims the GPS and navigates by skylighting the land. The cabin adrift, a warm speck in the night, red cherries of cigarettes, the throb of pop above the engine. They will be back in two hours, home in four, at work in eight.

When they dock, Tyrone will crane the totes from the deck of the tender to the deck of the processing barge, which is moored permanently alongside shore. From there two men will emerge and wheel the totes inside the refrigerated hull. It is icy, the machinery is deafening, its gears reverberating through the hull. The salmon are loaded to the slime line. Each fish has a spike placed through its gills to align it for beheading, a split made down the belly by hand, the egg sacs set to one side, and a vacuum cleaner hoovers up the guts. They get an initial grading, discarding the too small, the ones with bites taken out of them by seals and other battle scars. Then they send them across to the factory.

Young women and men, in blue aprons that reach to their boots and with sleeves that reach to their gloves, in hairnets, stand facing each other across the conveyor belts. The salmon, headless, gutless, pour into the factory down stainless-steel chutes in torrents of ice, sparkling under the factory's strip lights like a brook caught in the sun. The fish go down the line. The fins are sliced away. Fins litter the floor around the workers' boots, pathetic when divested of their bodies. The shorn fish are passed to the man at the splitter, and they come out in two halves. They continue on down the belts where a team with knives trim off any excess or ugliness. The fillets glide through the pin-boner, which epilates each piece of flesh to extract the tiny bones, and arrive at a final quality control. From here they will be packed, and sent to the freezers.

The fish leave Emmonak in freezers in the hulls of the barges that bring in most of the town's supplies. They ship south to Dutch Harbour, where they are repacked into containers, and from there they will go to Seattle. From Seattle they are loaded onto great container ships and fan out across the globe. Some begin a journey down the Pacific Coast, through the Panama Canal, and across the Atlantic, to arrive a month later at an English port, Felixstowe perhaps, or Immingham. They are held in a central warehouse in Grimsby, and from Grimsby lorries take them down the M1, to arrive at New England Seafood at Chessington in Surrey. The fish have journeyed perhaps thirteen thousand miles, a little more than they might have travelled in their lifetimes. Months later, in the winter, back in England, I go to visit.

The smell is the same, but that is all that is the same. At the end of an industrial estate in southwest London, near Heathrow, just off of the M25, past Screwfix, past Ash and Lacy Building Systems, past the recycling plant, is a picture of a tuna against a sepia backdrop of a globe. Behind the fence are the anonymous, cubic warehouses of New England Seafood. Articulated lorries tick over in the forecourt. I am approached by security, let through the turnstile, and asked a variety of questions: Do I have dysentery, typhoid, tuberculosis? Have I been to a slaughterhouse in the past twenty-four hours? Do I have any nuts on my person? Planes pass overhead, one every minute. I am given a lanyard with my name and purpose, and I am escorted to reception.

Max, in his thirties, wears a shirt and casual sweater, his hair held back by his sunglasses, a rubber band around his wrist that is connected to the internet and does something for his health. He glows with health. He does not like London. He spends his weekends in the country when he can, shooting, maybe, or fishing. New England Seafood was founded by his uncle in typically Alaskan fashion (a love of fishing, a trip to Nova Scotia, some dabbling in the fur trade, a nascent business importing live lobsters, storing them in bathtubs behind a Chinese takeaway). Their main focus is tuna, bass, bream and salmon. Max takes care of the salmon.

New England Seafood control 90 per cent of the wild salmon market in Britain. Their fish come from eight different catch sites, of which six are in Alaska. We go into the processing plant. I am given rubber boots, a fleece and a white coat. I must remove my watch, put on a disposable balaclava and a hairnet. We walk through a contraption that scrubs our boots, and then we wash our hands, and then we sterilize them. I am divested of my pencil, in case I leave it in a fish. The floor stands empty, the machines at rest. The workers, about five hundred of them, mainly Eastern Europeans, are currently on break. In the far room, one man is unloading fillets from a box onto a trolley, in preparation for defrosting. The box has come from Kwik'pak.

'It was like going back in time,' says Max of his first trip to Emmonak, in 2013. 'I loved it, and I hated it. But mostly loved it, I guess.'

New England Seafood visit regularly with each of their suppliers to police the level of quality that is required by British supermarkets. The Yukon remains the toughest.

'We'd have to try and make them understand that if a fish

falls on the floor, it's no longer fit for human consumption,' Max says, incredulous that anyone could think otherwise. You should see the smokehouses, I think. But then, I have met Athabascans who believe that anything less than a king is fit for consumption only by sled dogs, so I guess we all have our standards. Max found no line managers at Kwik'pak, no technicians. A fisherman's careless handling, a misplaced knife, a blunt machine, freezing too soon or too late: anything can impact the quality of a fillet. Wrong labelling can cost millions.

'Their work ethic,' he says, 'was appalling.'

Things are improving now, both the quantity and quality. The first year they shipped eight containers, this year they did sixteen.

'But we still aren't where we'd like to be,' Max says.

I pick up a packet of two frozen fillets, with the same Kwik'pak logo, of a Yup'ik man in a parka, holding up a salmon. I have rarely had such a tangible grasp of the nature and scope of globalization. Keta salmon, it says, which is how they market chum here. *Oncorhynchus keta*. There is a freeze date (25 June, a couple of months before I arrived in Emmonak), a catch method, a catch site number, a traceability number. Each fish will be defrosted, unpacked, and checked once again for quality. Bruises, blemishes, blood spots, gaping flesh, soft spots, belly membrane, exposed skin.

'It's like a carrot that's the wrong shape,' says Max. 'It doesn't really fly.'

They will be trimmed, if necessary, and chopped into chunks, and a machine will weigh out two pieces that comprise 220 grams. Offcuts go into packs for stir-fries. Another machine vacuum packs them, and another machine adds the requisite label. New England Seafood supply all the major supermarkets: the same fish, same pack, different labels communicating different values, with vastly different prices. Lorries will take them away this evening. They will last six days on the shelves. Then they go in the bin.

I go to buy some. I go to the Tesco on Seven Sisters Road. By now it is impossible to know where it is from – supermarkets assume consumers have little interest in whether this salmon was intending to swim up the Yukon or the Kobuk. Most supermarkets don't even bother with Alaska as a qualifier, preferring, instead, simply 'wild' or 'wild Pacific'. Wild Pacific salmon could also be Russian or Canadian, but 'wild Russian salmon' does not hold the same mystique.

I find them in the chilled aisles side by side with other fish, beside farmed salmon that have spent their lives in cages, beside other wild fish that have lived other lives in other seas. It costs me £3.50 for 220 grams, which means that, if this is chum, then the price has increased elevenfold since I first saw fishermen on the tender paid seventy-five cents a pound for it. It doesn't seem much, considering the journey that it's been on, the number of people who have conspired to move it from the Yukon to my shopping basket. It has one final journey. I cycle home. It is four o'clock and already dark. I fry it up with some rice. A splash of soy, a pinch of salt. There are alternative ways of cooking now that I am out of the bush – a woman I know in Whitehorse likes to poach her salmon in the dishwasher, set to the sanitize cycle – but I want to keep it simple.

This is as close as I'm now going to get to one that has been pulled out of the river, and it is very nice indeed. It is perhaps a bit more handled, a little slower to be gutted, but this fish is not profoundly different from those that an Alaskan will pull from their deep freeze out of season. The size is a surprise, of course. I have been used to gorging, to eating by the handful, to making my way through several pounds of flesh that have no economic value. This portion is not a handful. But it is flakey and fatty, and I think I can tell a difference from the farmed fish, although out of context it is hard to be sure. And it isn't king, of course. I eat it looking round my kitchen, listening to the news. The taste of a fish still dripping with the Yukon, charred over a fire and handed to you on a paper plate, creased in the middle by the weight of it, cannot be overestimated.

There are ten more miles to paddle. We take them slowly. We are in no hurry to arrive. Kwiguk Arm is one of the narrower runs to the sea; the Yukon, braiding out across the delta, reaches it by a multitude of means. Gulls and ravens tumble against a grey and windy day, but there is, at least, a break in the rain. Two juvenile golden eagles perch on a pile of drift, mantling, their teenage feathers mottled and scruffy. A burst of ducks takes flight. We drift, and we listen to the unoiled whistle of their wings. Above the clouds, a jet plane.

We moor up on the near side of one of the many islands in the mouth. All about is flat, immensely flat. On the beach, in the lee, the day is still and almost warm, a thin, high warmth, but when we climb the bank to the flat of the land the wind seizes hold of us. We bend to it. Since Russia we may be the first objects to have stood in its way. Leonine grasses arch away from the ocean, and except for one copse of scrubby alder in a divot in the landscape, there is nothing else at all. We step through the grasses, up to our knees, and the whole land is burnished by the colours of autumn. The wind pricks tears from my eyes. A high crackle of birdsong. I feel euphoric, just to be here, just that this is real.

In the middle of the island is some sort of beacon to warn passing ships of land, and it is impossible to gauge its height because all perspective has long since vanished. It seems incredible that any ship should pass out here, that this place could be on any other journey but our own. If there was a place to topple off the end of the world this would be it, where the map falls to tatters. It is as though the world has run out of ideas, and is now simply petering out, a scatter of islands and mudbanks that blur into the water, some grassy flats and a slop of mud and then nothing but the ocean. The light is fierce and clear and wild, the colours whipped up as before a storm. The grasses swell and ebb in echoes of the ocean.

These islands are the tops of other mountains. Everything here is so bare, so undisguised, that the bones of geological processes seem more comprehensible, stripped back to their essentials, everything so slow and spacious that one can observe at the speed at which the world works. And now, on the horizon, it is water and not land, the first horizon not hemmed by banks that we have seen since the trip began. It is the Pacific Ocean, the Bering Sea, and we have paddled two thousand miles to be here.

We quicken our pace. The sea pulls at us as the moon tugs on the tides. And as the horizon opens out and the land falls away, the ocean ranges vast before us. A flock of murre scoot low and long above the line of water. No cliffs or rises, no islands far out, no passing tankers, nothing but the end. It is so flat it appears that on a wing and a prayer we'd have a decent chance of paddling to Russia.

We throw off our clothes and dive in. It is compulsion, some way to mark the end. In a month or two it will be entombed in ice, and yet today, 6 September, it is not particularly cold, no colder than the river. The water still tastes somewhat sweet. There is so much water barrelling down these estuaries, 200,000 cubic feet per second, that it will be sweet for many miles hence. And it is silty still, the sea a pale and milky grey, and our pale bodies loom beneath it. There is a blue line far, far out, where the salt and tide take over.

We get out, dry off, and sit there, looking out. Right now salmon smolts will be tumbling from these river mouths, bound for their feeding grounds, untethering from the land and hurling themselves into another world, one that is saline and entirely without boundary. But here we must stop. We can follow only so far.

A gull lifts from the shore, is pulled out to the horizon by the wind.

A week later and we are back in Whitehorse. We sit in a café on Main Street eating expensive salads and avocados on sourdough toast. Our waitress brings us smoothies made of vegetables. Her arms have tattoos of forest scenes. On the chalkboard there are twelve different types of coffee, there are cinnamon buns the size of saucers. My hair is cut, my trousers are new. It is still warm here, as though the seasons, too, were carried off by the river. The Yukon that runs through town here looks small and crystal and benign. Every other vehicle seems to be driving around with a canoe on the roof, upturned, like a stetson. In a little while we will walk back to Hector's house, where we are staying. Maybe we will put a record on, or maybe take a bath. There will be no bears, or storms. We packed our gear inside the canoe and flew the whole lot to Fairbanks. We flew ourselves to St Mary's, changed planes and then flew to Anchorage. We checked into a motel, got roaring drunk by our second beer, and inexplicably ordered salmon in a restaurant. The next morning we took another plane to Fairbanks, met the canoe, and sold it. And then some friends drove us to Whitehorse. We are back in Canada, and our flight home leaves in less than a week.

We spend a day at Takhini hot springs, scrubbing off the grime. We spend another day at the Yukon Wildlife Preserve, looking at lynx and mountain goats behind the bars. At the dam, in the visitor centre, you can look through glass into the holding pen at the summit of the fish ladder. The guide, a girl from Teslin on her summer job, tells us that they had a single Chinook pass through yesterday, but that there have been none today. There are no other tourists. Even here, at the Yukon's source, summer is at its end.

We head up to the hatchery to see how my five thousand eggs are doing. They're doing fine, I needn't have worried. Lawrence Vano, one of the two managers, is still there, in the office by himself; there is not much to be done this time of year. Most of the smolts have been released. I ask him how the season has been. The returns through the fish ladder are about what was expected, a little up on the 1,500 average, but the female to male ratio is way down, less than 18 per cent. It could, of course, be a blip; Lawrence's thirty years here have been enough to weather some changes. One year just 162 fish came back. But Lawrence finds the lack of females a cause for concern.

As always, it is impossible to know; data sets, opinions, cultures, conflict. The year 2016 saw Chinook numbers entering the river at the upper end of preseason estimates, and higher than the average for the previous five years, although numbers are still way down on what they once were historically. 72,300 fish crossed the border, almost 20,000 more than required by the Agreement. These increases are thought to have been achieved by conservative management, and in consequence many fishermen reported not meeting their subsistence goals because of the lack of fishing opportunities. There is a cautious optimism, although one laden with caveats. 'If we have a weaker than expected return of females or a really skewed age structure in the run, that is always a bit of a flag for us to say perhaps we should not get too excited just about the numbers,' said Mary Ellen Jarvis of the Department of Fisheries and Oceans.

It may never be agreed what caused the crash in the runs of the kings. But we know where we are now: strict management seems to be helping to get salmon onto the spawning grounds, but those salmon are still small, they lay fewer eggs, and this means that the productivity is not far off one to one. The system, if not imperilled, is extremely precarious. 'There is an opportunity now to repopulate those fish camps,' Stephanie Quinn-Davidson, ADF&G's Chinook manager up until 2015, has said to me. 'There is an opportunity to try and get people back out there and pass on the culture and tradition. But I think people are scared that we're not in the clear.'

Salmon can be brought back from the brink. They are swimming through Sheffield in England since the rivers have been cleaned, after a gap of two hundred years. They are back in Portland, Oregon, and Paris, France. There are not many, but it is a start. And the future of the Yukon king does not seem hopeless, either, if reconciliation can be achieved. A reconciliation between commercial and subsistence, between those who live at the mouth and those who live at the source, between those who see their entitlement to food and wealth and culture swimming past them up the river, and those who want a conservative approach, if not an outright ban, forever. The life of a fish and a river, as I have learned, is astonishingly complex.

We are part of the landscape, and the salmon's story is our story. They are the lifeblood of this land, coursing through its veins, and they are the lifeblood of the cultures still connected to this land, who are fighting to determine what their future looks like as the century unfolds. A genuine approach to managing the king can only be holistic: the creation of an ecological web that is able to integrate also culture and politics, to integrate histories and stories and beliefs. Nature is not something else, isolated, out there; it is as much a part of us as we are of it, and neither can be altered without impacting on the whole. Whatever we choose to do, we cannot pretend that we did not know. What is certain is that, for the king at least, the Yukon is the last chance to get it right.

★

A few smolts remain at the hatchery, left over from some school project, and Lawrence asks us if we would like to release them in one of the tributaries upriver. He rests his hands on the edge of the tank and peers down at them with a faint smile, as though observing his own children from the touchline. He picks up his net and dips it in, jabbing it about like at some fairground sideshow.

'The hard part is not killing them,' he says.

He catches the first ones easily enough, but the last few are more wily, putting up a spirited chase about the tank, little tails vibrating frantically. We end up with thirty-one in two buckets. I take one, and Ulli takes the other, and Lawrence follows us out to the car. We put the buckets on the backseat and fasten them into seatbelts. Lawrence looks on, and I worry, for one ludicrous moment, whether he might be about to cry.

We drive out of town, up out past the airport and towards Carcross. Carcross, where one of the Indian boarding schools once was, before it was burnt to the ground. Distant mountains that will become the Rockies further south open up before us. It is satisfying to be moving again after several days inside. We could just keep on going, I think. This road goes to California. I watch the buckets in the rear-view mirror. The roads are wide and spacious, a few trucks drifting down them. Some miles from town we take the turning for Wolf Creek, a narrow stream bound for the Yukon. It is midweek, a little after noon, and we have the whole place to ourselves. Some signage discusses the life cycle of the salmon. The creek pops from a culvert beneath the highway and runs the boundary of the carpark before turning into the trees. We take the buckets from the car and carry them down into the woods. I have a look about. I can't shake the feeling that I am somehow up to no good.

It is cold beneath the trees, and Wolf Creek is slatey in the shadows, a dozen feet across. It rushes over low smooth rocks, bearing the first of the autumn leaves along. A robin hops about in the brush. I pop the lid from one of the buckets, and watch the salmon, turning circles. Their speckles are far more prominent than in their adult counterparts, their small features fixed and determined. I wonder about doing this. It seems like one more mad example of how we have locked ourselves into an alien ecosystem, the reproduction of another species now dependent on ourselves. The whooping crane fledglings raised by humans that must be flown by plane because they do not know how to migrate; the apple trees in China that must now be pollinated by hand. And yet, I will remember these salmon. And so will the children who raised these ones as part of their school project, who might watch in a few years' time as they scale the fish ladder on their return. Maybe locked into the ecosystem is where we need to be, not as God, but as compatriot; just where the stories from here have always had us.

We find a spot where the creek has pooled, more clement than the rest. Some pond skaters skim the water. I lift my bucket, and Ulli lifts hers. I feel like we should say something, to bestow a little ritual on the moment, but I don't know what to say. So we just pour them in. The water spills and they huddle in the shrinking space and then, out of options, they tumble forth. The world becomes bigger than they ever could have known. There is no transition, no moment for contemplation. They are gone, swept off and away downstream. That can't be right. I panic. I jog along at the current's pace. I think I see one for a moment, further down, riding out a rapid, and then the sound of the rush of the water is all there is, and there is nothing else at all.

I walk back up the bank, to Ulli, and we stand there, beside our empty buckets. In five years, or six, we might see one of them again, if they are very lucky. In a few minutes they will be catapulted through the turbines of the dam. That will be a third of them gone, and that is just the beginning. Just 0.03 of these fish have a hope of making it back alive. We drive back to Whitehorse for a coffee.

There is one more trip to make. We drive north, the Klondike Highway. Down north, as the locals say, travelling the direction of the river. This is the road to Dawson, and in a car it takes six hours. It is two weeks in a canoe. The trees are flame yellow, as primary as the sky. Families are parked up at lakes for picnics, barbecuing meat. On the radio it is rock or it is country. We listen until we are out of range. There is a steady stream of RVs going the other way, bound for winter pastures before the roads up here freeze over. The geese, still moving south, keep pace above them.

The road unrolls, tracking the river, echoing its shape, leaving it for a time, rejoining it, both finding their paths of least resistance through the landscape. A cabin here, fifty miles, a cabin there. Three hours north of Whitehorse is Carmacks, population 500, notable only for being a place on a road otherwise empty for three hundred or so miles, and memorable for a burger that I had here the first time I paddled through. Half an hour north of Carmacks the road dips to a valley, and here there is a turning for the campsite at Tatchun Creek. We pull off the road and turn in. There are a few cars and a single RV, spread out through the trees, their fires glowing in the dusk. It is companionable, to camp with others. There are dogs in the RV and they worry at the darkness.

Tatchun, from the old Athabascan for 'fish back'. The creek runs swift and shallow from out of Tatchun Lake, and joins the Yukon a few hundred metres below the campsite. We walk down the path to the confluence. I remember passing here, watching this mouth as I floated past. There are two canoes out there now. This was once an important fish camp and trading place for the many different tribes in the vicinity, as a cracked and moulding information board informs us. Microblades found in the area date back 7,000 years. '*Gyò* means Chinook salmon', says the board. '*Tatchun* means fish back. *Tachän Gyò Huch'än* means see salmon's back in shallow water people.' There is nothing, now, but a parked-up van skipping the twelve-dollar fee for the campground, a fish rack that may be either a reconstruction or a relic, and some few remnants of a night spent drinking, smashed glass, cigarette butts, charred tinfoil in old embers.

We have come to look for kings. I am excited. Despite four months of paddling, of talking about every aspect of their lives, four months of cooking them and eating them and pulling them from nets, of watching as they are disembowelled, gasping, their hearts still twitching, and are hung from racks to air, despite all of this I have yet to see a truly wild adult king. By which I mean one that is bound for something other than human interaction and consumption. I have seen blips on sonar, I have seen photos on iPhones and cans of strips and strips of jerky. And although this is a story, above all else, of relationships, of the symbiosis of people and fish, of the imprint that one leaves on the other, there is a part of me that cannot help but root for the others, the ones that make it through, ungilled, unseen, and that return. Because of them, the others.

The Tatchun gets some of the very last spawning stock of the season. Al von Finster, a retired salmon biologist for Fisheries and Oceans, had emailed me his advice, and it had instilled in me a thrilling sense of imminence and panic. 'You'll be at the end of the run,' he wrote, 'and it can end VERY quickly. One day fish, next day none. A day can make all the difference.' It was September and we were pushing it; Al had seen some a week ago. Every day that we lingered in Whitehorse my anxiety increased. 'Take a heavy gun or bear spray,' he wrote, 'and don't go at dusk, night or dawn.'

The next morning, well after dawn, we set off up the creek.

The air is chill, shot through with sunlight. The banks are high, a knot of tree and shrub and plant, and every bend is blind. The leaves are falling ceaselessly, catching on the water's surface. 'Hey, bear,' we shout before every turn, so as not to startle them. We could shout anything, of course, but that seems right. The hope is that if there are still bears here, and still salmon, then the bears will be much more focused on them, with their rich and oily proteins, than on our supermarket produce, or ourselves. The night before we had gone looking for bears, driving around like kids up to something illicit, emboldened by the car. We hadn't found any, and I felt both disappointment and relief. They like to feed at dusk when the salmon are in spate, and we had sat on the dirt road that ran the length of the bluff above the creek, watching the sunset, waiting. A kingfisher hammered a series of straight trajectories at the surface of the water, and emerged, at last, with a fish. But the bears did not come.

The leaves of the fireweed that cover the banks are red and rusty now; some with a single flower or a couple at their apex, but most with only a puff of seeds like cotton candy dusting their otherwise bare stems. Summer's hourglass is at its end. There was a suggestion of frost on the tent again this morning. More disorientating than the endless light of summer is the speed with which the autumn now takes hold, as though launching an assault, the summer's rapid senescence. Tonight there will be seven more minutes of darkness than last night. The temperature drops daily. Rosehips catch the low light, already bletting on their branches. It could snow, any day.

The creek is only thigh deep at its deepest, but the current is swift and the water is cold and the riverbed loose and slick. We lean on branches we have cut and haul ourselves upriver, like weary travellers in a pantomime. On the outsides of bends where the water runs fastest we link arms and anchor ourselves to one other, our footsteps choreographed. We cross from one gravel bar to another, and plunge to the creek again. I am looking for redds, lighter patches of gravel, and the whole premise seems suddenly absurd. To be walking upstream where the bears are maybe feeding, stomping through a river in the hope of seeing skittish fish that in all probability have vanished days ago. Fish that are meant to have travelled 1,500 miles from the ocean to my very feet. The absurdity of their life cycle seems manifest right now. How could they *possibly* be here? *Right here*? At a distance, in the campground, I hear the RV drive away.

And then I see one.

A bolt of red slides across the creek bed. It is a female, alerted by our clumsy presence, and she abandons her redd and slips upstream into a slough. She is indistinct beneath the surface, but she is unmistakable. She is not a big fish, but she is here. We stop and sit astride a snag midstream, our feet out of the water, and silently, we watch. They are *here*. We have narrowed an ocean to a bend in a creek, and this is where they are. We wait.

At length, she noses back into the flow, tacks to face it, and drifts back to her redd, where she holds, balancing herself against the current. She hangs there, gasping at the water. She is the one in ten thousand that has made it. I wonder if she has some sense of the odds she has defied. Beneath the water she shimmers with the quality of stained glass, flexing with the current. There is still determination to her movements. For months they have fought the current, but the fight now is a fair one. For months it has been their constant, as gravity is ours, and as our perpetual fight is so instinctive that we are unaware of it, so they too must be unaware of theirs until they start to falter. Presumably they eye us through the water's surface, and wonder how we spend all of our lives balancing on two feet. What else manifests our ageing process more than the first fall, the broken hip? And this fish, with her flaccid body that no longer does her bidding, tries to hold herself steady in the flow and keep her offspring close.

But some anomaly of current catches her, and her body has lost its instinct or its strength or its capacity to act. The current takes her, tumbling her down through the creek, the sea exerting its pull once more. She rights herself, recovers her poise and ploughs back into it. As though swimming up through mud, she slogs back to the redd.

She will succumb. The river's eagerness for the sea will prove too much for her, and she will turn belly up and let go. She will slide down the creek, wheeling over the rocks, and flop out into the Yukon, or be pulled from the water, decomposing, and break down into the land, scumbled amongst the mushrooms and the leaf mulch. The season will advance without her. The ice will begin to creep. By the time that the first snows come she will be little more than backbone.

We carry on. At each bend I feel like we're pushing our luck. It is a palpable fear of the unknown, such as I have not felt since a kid, when monsters were real. Every time we have let those instincts get the better of us, but this time we override them and press on. On the banks there are salmon with their bellies chewed out. We see two more kings beneath the water, side by side, perhaps a pair. Another bend. And now the creek braids, and there are tributaries pouring in from all sides, rushing in through briars and flowing around tussocks. Beds of grass bow to the current, dun coloured and dead beneath the surface. The banks step back, and the stream becomes spacious, with broad views of the sky. Many of the nearby trees have fallen, jack-strawed by the wind to improbable angles, borne up by the others. Beyond, the water floods into the spruce forest, and all of it is dead, its roots drowned, needles still on the branches as though petrified in an instant. A bald eagle sits at the top of one tree, observing. Sticks and logs in four-foot lengths have been piled on their end in an undulating curve that crosses almost the whole width of the valley, and beyond the beaver dam the lake is deep enough to be black and bottomless, with hanks of weed waving in the gentle breeze of the current. In these months of travelling I have seen little that has altered the landscape so profoundly. Some few salmon mill in the darkness here. There are more dead kings on the sandbar, rotting down into the earth.

The water is too deep to carry on. I am nervous, my skin is crawling, I am ready to turn back. We take a final look around. The sky, and the forest, and the river, and the fish. The last of the year's swallows blow away in gusts over the mountains to the south, like summer's embers.

And the eggs will wait for spring beneath the riverbed, all orphaned.

Acknowledgements

It turns out that going to one of the least densely populated places on the planet requires a lot of help from a lot of people.

Many thanks to the companies that provided us with free gear for the trip: to Klean Kanteen for the drinking bottles; to Ocean Signal for the locator beacons; to Smidge for the bug dope; Varg for much of the clothing; and to Frost River for a particularly luxurious tent. Kanoe People and Up North, both outfitters in Whitehorse, were very generous with their time and with their kit. MSR, Big Agnes, Peak UK, Overboard and the map room at Stanfords were also a great help in enabling the budget to stretch further than it otherwise would have.

There were a huge number of people that I met along the river who shared their stories, their thoughts, their time, their hospitality and their fish with me, far more people than I was able to include in this book. There are many whose names I never knew, but thanks to: Stephanie Quinn-Davidson, Mary-Ellen Jarvis, Jan Conitz, Warren Kapaniuk, Lawrence Vano, Jesse Trerice, David Neufeld, Sandy Johnston, Chris Stark, Todd Brinkman, Karen Dunmall, Reed Morisky, Virgil Umphenhour, Scott and Robbie MacManus, Robert Neu, Gillian Rourke, the Teslin Tlingit First Nation, especially Bert Goodwin, Duane Aucoin, Madeleine Jackson, Bob Magill, Sandy Smarch, and Richard Dewhurst – the Tr'ondëk Hwëch'in First Nation, especially Roberta Joseph, Peggy Kormendy, Debbie Nagano, Percy Henry, Angie Joseph-Rear, and Julia Morbery – Jody Beaumont, Tim Gerberding, Cor Guimand, Natasha Ayoub, Emmanuelle Gauthier, Kate Rorke, Andy Bassich, Rhonda Pitka, Cliff Adams (who tragically died on the river in May 2017, along with his wife Ai Nakjima-Adams), Paul Williams Sr, Richard Carroll, Mary Jackson and family, Walter Chuck, Andrew Firmin, Joe, Jessica and Jason Burgess, Charlie Wright and family, Stan Zuray, Steve O'Brien, the Campbells, Don and Annie Honnea, Delbert Mitchell III, Joey Katches, and everyone else at the Kokrine Hills Bible Camp, Isky and Ed Sartin, Jenny Pelkola, Amy Graham, Ellie Stickman, Richard Burnham, Mike Maillelle, Tawnya Richardson, Kathy Chase, Connie, Bergen and Mary Demientieff, and Mary Junior, Kyle Schumann and everyone at the Pilot Station Sonar (thanks for teaching us crib, Dave), Donn Gallon, Janice Kelly, John Tinker and family, Jack Schultheis, Jim Freidman, Judi Murdock and everyone at Kwik'pak, Lindsey, Tyrone and John, Max Ropner and New England Seafood, Jim Davis, Wouter Redelinghuys, Jenn Round, Mike Williams, Felix Flynn, Michael Cresswell, Ivan Ivan, Noah Okoviak, and Samuel Jackson. Thanks also to those who helped us prepare for the trip before we left, to Jim Hamer, Dom Kawalec, Kirsty Porter, John Lengthorn and Rob Davis, and to Jock Barge for some particularly memorable sea-kayaking.

In First Nation and Native Alaskan cultures, stories have traditionally been passed down orally, and have only started to be written down comparatively recently. Everyone I met was aware that I was writing about the Chinook, and they chose to share their stories with me. Thank you. The string stories, and the story of the black fish, were told to me by Grant Kashatok, in Newtok, in 2013. The story of the Salmon Boy has many sources, but this telling is taken from *Monuments in Cedar* by Edward L. Keithahn. The story of Raven and the river flowing both ways comes from *Make Prayers to the Raven* by Richard K. Nelson.

It was not the intention to produce a fully referenced text, but there are some works that I relied upon more heavily than others, either for their information or for the ideas that shaped the thinking behind this journey: *Salmon* by Peter Coates, *The Other Side of Eden* by Hugh Brody, *Coming into the Country* by John McPhee, *Yuuyaraq: The Way of the Human Being* by Harold Napoleon, *Finding Our Way Home* by the K'änächá Group, *The Fall of the Yukon Kings* by Dan O'Neill, *The Last Breakup* by Ted Genoways, *Carry On* by Tim Attewell, and *Arctic Dreams* by Barry Lopez. *Paddling the Yukon River and its Tributaries* by Dan Maclean, and the maps of Mike and Gillian Rourke, got us safely from one end of the river to the other.

To all of you who read bits or whole or multiple drafts, and put up with a lot of salmon chat along the way: Effie, Dan, Erin, Gerry, James, Gareth, Harry, Mike, Pete, and Mum and Dad. To Pete Westley and Al von Finster for patiently answering as many questions on salmon as I could come up with. Thanks to the Winston Churchill Memorial Trust, who funded the first trip to Alaska back in 2013. My thanks to Jenny Rothenberg Gritz, then at the *Atlantic*, who ran the original article about the trial on the Kuskokwim. And thank you to the team at *Lacuna*, in particular Andrew Williams and James Harrison, who published many of my original Alaskan articles, introduced me to my agent, and generally got the whole thing off the ground.

Thank you to a wonderful team of publishers: Helen Conford, Vanessa Mobley, and Amanda Betts – your perceptive and patient work has turned this into something far richer than it otherwise would have been. I have hugely appreciated your time and dedication as I've figured out how to write a book. Thanks also to Margaret Stead, Shoaib Rokadiya, Richard Duguid, and Richard Mason, and to the teams at Allen Lane, Little Brown, and Knopf Canada for pulling everything together and guiding me through the whole process. Thanks also to the team at C + W, in particular to Emma Finn, and a massive thank you to Sophie Lambert, who spotted a nascent idea a few years ago and helped to grow it into this. I could not wish for a better or a more supportive agent.

Some people went far beyond the call of duty to help make this journey happen. Thank you Beth Hamer, for all those canoe trips and for all those conversations – I'm sorry how things worked out and I look forward to new beginnings. Thank you to Ben Stevens, Orville Huntington, and Ann MacKenzie for driving canoes and people and gear all over Canada and Alaska. To Peter Mather, for much help as I began putting plans together, and for an unforgettable day on the Takhini. To Gale and Erik Vick, for inviting us into your home and having us there for so long that you must have wondered if we'd ever leave. It was exactly what we needed after four months in a tent. This would have been a much harder trip to make, Gale, had it not been for your help. And to Hector and Miche, for wonderful meals and long talks – thank you for making Whitehorse feel like a home. Your kindness, generosity, and hospitality were massively appreciated, and I look forward to returning them one day.

And lastly, to Ulli. For the company, for the memories, and for catching all the fish. For all the nights you did the work around the camp so that I could write, and for your unerring support during the planning, the research, and the writing of this book. It would have been a very different thing without you. I'm so happy to have you paddling with me in the boat.